Rivers under Siege

Rivers under Siege

The Troubled Saga of West Tennessee's Wetlands

Jim W. Johnson

Outdoor Tennessee Series • Jim Casada, Series Editor

The University of Tennessee Press / Knoxville

The Outdoor Tennessee Series covers a wide range of topics of interest to the general reader, including titles on the flora and fauna, the varied recreational activities, and the rich history of outdoor Tennessee. With a keen appreciation of the importance of protecting our state's natural resources and beauty, the University of Tennessee Press intends the series to emphasize environmental awareness and conservation.

Library of Congress Cataloging-in-Publication Data

Johnson, Jim W., 1940–
Rivers under siege : the troubled saga of West Tennessee's wetlands / Jim W. Johnson. — 1st ed.
    p. cm. — (Outdoor Tennessee series)
Includes bibliographical references.
ISBN-13: 978-1-57233-490-8 (alk. paper)
ISBN-10: 1-57233-490-8 (alk. paper)
1. Stream restoration—Tennessee, West.
2. Rivers—Tennessee, West.
3. Nature—Effect of human beings on—Tennessee, West.
I. Title.

QH76.5.T2J64 2007
333.91'6215309768—dc22   2006021750

To Bennett and Mildred Johnson, my parents

Contents

Illustrations

Figures

Maps

Foreword

One morning a decade ago, as I indulged in my daily ritual of perusing the newspaper while enjoying oatmeal and orange juice, a headline that would appear with increasing frequency in the coming years caught my eye. The news story told of an interstate catfight over water—one that involved a devil's brew of politics, socioeconomic factors, and environmental concerns. The stream in question did not fall within the geographic boundaries of this book, but it might just as well have. After all, water is earth's life blood no matter where one lives, and as human populations explode, that fact is being driven home in increasingly telling fashion.

We, as humans, have shown an all-too-common tendency to foul our waterways, and the pressures we put on them for all sorts of uses—drinking water, irrigation, industrial purposes, a means of carrying away sewage and other effluents, recreation, and more—have increased exponentially. In the final analysis, as the title *Rivers under Siege* suggests, such considerations form the subject matter of this book.

As Jim Johnson makes clear in a work that combines solid research with a passionate love for the streams he covers, the residents of West Tennessee, whether they know it or not, have ample reason for concern. Indeed, in the case of some streams, outright alarm might be a more appropriate reaction.

Johnson opens with an overview of the watersheds—the Wolf, Loosahatchie, Hatchie, Obion, and Forked Deer—then launches straightaway into a riveting, wide-ranging discussion of their history and the story of those who have used and abused the streams. Reelfoot Lake, a unique place, also figures prominently in the picture. Through Johnson's carefully documented, insightful, and at times deeply troubling treatment, we see it all—the good, the bad, and the downright ugly. There are, alas, culprits in abundance to be encountered throughout these pages. They include self-serving politicos, narrow-minded and wrongheaded bureaucrats, agricultural exploiters, misguided wildlife and wetlands managers, and more. Set against this parade of, dare one say it, shady or, in some cases, downright sleazy characters, are a host of good guys. Their numbers include caring sportsmen, landowners whose perspective extends far beyond a quest for the almighty dollar, and "little guys" within cumbersome bureaucratic labyrinths laboring quietly but valiantly to do things right. Committed, even visionary, individuals of this sort provide glimmers of brightness in these pages.

That is just as well, because in the final analysis *Rivers under Siege* is a chronicle with more than a fair share of doom and gloom. If you come away from reading this cautionary tale without damning those destructive builders of dams and crazy channelers of rivers, the Corps of Engineers, suffice it to say that you are a singularly forgiving and charitable soul. Yet the corps is by no means the only readily identifiable culprit here. Historically, the U.S. Fish and Wildlife Service has not always performed the way it should, and anyone who has followed the sometimes mindless meanderings of state leaders in Nashville will not be surprised by shortcomings aplenty from that quarter. Even those who might seem to have the greatest vested interest in "doing things right"—fishermen, duck hunters, wetlands managers, biologists, and landowners—have not always performed in acceptable, much less admirable, fashion.

Love them (which would require a saint) or loathe them, irritating personalities and irresponsible government entities form an integral part of this tale of five rivers. It is a tale told in compelling fashion by an individual who has been personally involved in stewardship efforts. In that sense, *Rivers under Siege* is a cri de coeur, and notwithstanding the somewhat upbeat stance Jim Johnson takes in his conclusion to the book, these storied streams are on the brink of an environmental Armageddon.

Once they were waterways that wended and wound their way across the West Tennessee landscape in a fashion reminiscent of the laughter lines on an old man's face. Today a more appropriate analogy might be worry lines or even scars, as chapter titles such as "The Burial of the Rivers" and "The Rivers Held Hostage" make abundantly manifest. I cannot imagine anyone reading this work and not leaving its pages deeply troubled.

Certainly this is not a "feel good" book; far from it. Nonetheless, every Tennessean, indeed anyone who cares about the natural world and its vital role in our existence, should read it and carefully ponder the meaning of this saga of five rivers and a lake. To do otherwise is to continue the ostrich-like rejection of reality that has too often been the official approach to the ongoing, shameful siege of these once wonderful waterways.

There have been a number of books in the Outdoor Tennessee Series that delved deeply into environmental matters. The message and contents of some of them, most notably Richard Bartlett's *Troubled Waters*, stirred my soul. The stream covered in Bartlett's book, the once terribly polluted Pigeon River in East Tennessee, has successfully taken the first steps along the long road to recovery, and today spunky smallmouth bass swim in eddies and hide amidst rocky riffles where once there was little life and a great deal of black, ugly water adorned by PCP-laced foam. One can only hope that this book will help engender the sort of concern that leads to comparable developments with rivers at the opposite end of the Volunteer State.

It is a book eminently suited to the Outdoor Tennessee Series, and as I look back to the words I wrote a decade ago in my foreword to *Troubled Waters,* the concluding paragraph to that introductory material is equally applicable to the present book, with only slight changes, such as the substitution of Johnson's five rivers for the Pigeon. Accordingly, I'll conclude with an adaptation of what I wrote then.

One thing is certain: whatever your perspective you cannot read this tale of troubled waters and remain unmoved. To me the story as it has unfolded thus far is tragic; yet I leave these pages with a fervent if muted hope (one shared by the author) that somehow, some way, someday, they will undergo a transformation that is magic. On that glad morning, and it needs to come soon or it will be too late, the five rivers will flow much as they once did, pure if not pristine, filled with fish and nourishing wildlife on their waters and in adjacent wetlands. For now, however, we have a sobering study of a shameful story. Read it and weep or wonder. Most of all, though, pause to ponder what this saga of streams and Reelfoot Lake tells about the difference between sensible use and sad abuse. As you do so, be constantly aware that it is water that sustains the good earth, which in turn sustains each of us.

Jim Casada
Series Editor

Acknowledgments

Rivers under Siege was written mainly at the encouragement of Dr. James "Jim" Byford, an educator and a friend to both the agriculture industry and the management of wildlife. He has never given up on the belief that there is a patch of common ground where controversy over the management and use of West Tennessee rivers can be resolved.

Why is it that the relationship between wetlands and people cannot be symbiotic? This is the central question that Dr. Byford and those of us who managed natural resources pondered as we faced the conflicts involving these rivers. To date, modifying the rivers has been disastrous to both the waterways and the people of the state. Our belief was that the situation could and should be changed. This book is an account of how a few dedicated conservationists tried to attain this seemingly unreachable goal.

Dr. Byford and the following persons provided materials and advice for this book, and to these generous contributors, I am immensely grateful. They include Ben Smith, a planner and diplomat who engendered consensus among the factions involved in the West Tennessee Tributaries Project, and Dr. Wintfred F. Smith, Reelfoot Lake historian, researcher, and colleague in the effort to better understand wetlands. To Professor Jack Grubaugh and Outdoor Tennessee Series editor Jim Casada, who reviewed the manuscript for the University of Tennessee Press, I owe thanks for their candid editorial comments and suggestions. Scot Danforth, UT Press acquisitions editor, and Gene Adair, UT Press manuscript editor, did much more than edit and give advice—they refreshed both my thoughts and the book itself. I am deeply grateful to them for this and for their patience and prompt replies to my endless questions. Frank Zerfoss, Don Orr, and Harold Hurst, dedicated friends and colleagues in this career-long struggle, took time to review the manuscript for accuracy.

Finally, I owe tremendous gratitude to the waterfowl managers, present and past, of the northwest territories. They include Carl Wirwa, Larry Armstrong, Paul Brown, and their crews—those still here to carry on our work—as well as Alan Peterson, Charlie Childress, Ralph Gray, Floyd "Speck" Hurt, and their crews—those who made their contributions and have gone. This story could not have been written without the dedication and contributions of these managers to the management of wildlife and wetlands.

For nearly thirty years we tried to change the way wetlands were managed along the Obion–Forked Deer rivers. Dr. Byford continually insisted during the last few years that someone should recount this history. As a witness and active participant in the river issues, I became his candidate of choice. Resisting until the end, I aimlessly drafted a few notes in October 2002, wondering about where to start and what good purpose it might serve. Dr. Byford helped clarify that question by saying, "Maybe at least those that follow us will not make the same mistakes."

I suppose that Dr. Byford's reasoning remains one of the primary justifications for this book. But it was also written to remind us that we should not (perhaps cannot) live without rivers as we envision how they should be, that riverine wetlands are more than mere mosquito-infested swamps and gutters to convey runoff and more than real estate to squander. The book also reminds us that we have yet to acknowledge the true value of rivers and that the status of rivers reveals the character of the country and of the people residing there. As Ernest F. Swift, author of *A Conservation Saga,* so eloquently put it, "Fat cities do not thrive on lean countryside." If this is the case, the citizens of West Tennessee cities have been on a sparse diet of countryside, since they have been deprived of one of their richest natural resources for nearly a century—native rivers and the wetlands.

Reviewers suggested that the first draft of the manuscript tended to limit the story to historic or professional interest. Jim Casada suggested adding more humanity to the story to encourage a wider readership; that request made for some difficult choices since this was a story filled with humanity. In addition, I was asked to better bridge the transitions from one part of the story to another, which was no small task since the events occurred over essentially the same period. I agreed completely with these suggestions, but then I overdid it and had to trim the manuscript back by about 25 percent.

This book does not dwell on the details of hard science to make its points; rather, it relies on practical firsthand accounts and interactions with the people involved. It is intended to serve wildlife biologists and managers, researchers of local history, administrators, and students. I hope, too—and this is no small consideration—that the book will be of interest to all citizens with an abiding interest in the restoration and future of wildlife and wetlands, especially the rivers that create and sustain them.

Introduction

As far as the eye could see, the river channel ran straight, unmistakable evidence that the Forked Deer had been reshaped by human hands. From a thousand feet on this day in 1985, the river corridor appeared to be a harmonious patchwork of green. Quiet pools, low-growing vegetation, and a pattern of drainage ditches and canals might have given one reason to think that special effort and purpose went into the design. From our Cessna 172, anyone who did not know better might imagine the river to be an oasis for wildlife, a coveted place for outdoor adventurers. Or it could be seen as an ingeniously designed landscape for agricultural purposes and economic prosperity. But on the ground there was a different story. The truth was that it was *not* an oasis, and certainly not a design for economic prosperity, but a river wrecked and ruined. Like a cleaver wielded across the back of a sun-bathed snake, the long, wide, and straight manmade ditch had disabled the former river. Eventually there would be scant evidence that the old river ever existed. What manner of mind could have caused this tragedy, and what useful purpose did it serve? The best we can answer is that it was probably arbitrary; mostly it was for convenience and economic expediency—for the present, for the short term, and not for the distant future.

After more than thirty years as a natural resource manager, I am still dumbfounded to find that we wasted what little was left of some of the richest and most sustainable natural resource banks on earth—native rivers. These rivers affected a million acres of Tennessee turf and more than two million of its citizens. You would think that economic incentive alone would have found the state clamoring for the best economists in the world to figure out the best way to manage this resource, but that did not happen.

Stretches like the one on the Forked Deer that we viewed from the air are common in the rivers of West Tennessee—among them the Obion, the Hatchie, the Loosahatchie, and the Wolf. Except for the Hatchie, there is little left of the old winding corridors on any of the rivers. Their forested floodplains now consist of life-deprived canals, stagnant swamps, and dead and dying trees. Soil erosion, levees and ditches, and arbitrary encroachment incompatible with river ecology marked the beginnings of the problem. Sediments, flooding, and the lack of drainage have, consequently, plagued farmers, landowners, and, indeed, the rest of us, as fish and wildlife, infrastructure, and all the pleasures we need and expect from river wetlands have been destroyed. This blight cannot be erased without aggressive and well-designed solutions and planning our footsteps on the landscape more carefully.

June 1985 was a time when we searched desperately to make sense out of the river crisis. Some were content that rivers like the Forked Deer were good enough—that they should be spared any further interventions. But a consensus on anything involving rivers was difficult to find when so few people understood them. For waterfowl managers like myself, it was a daunting picture; we had stared at it for years. How could we help fix it? The answers were slow to come but not as slow as changing minds enough to take action. While we waited, the dilemmas worsened, and the costs continued to rise.

We, the waterfowl managers in northwestern Tennessee, were in a position to see it all. The outlook was dismal for the future of waterfowl hunting. Maybe we could do something about it. We believed it was our duty to do so, but we also felt guilty because we had been part of the problem. Like so many others, we had made our mistakes with all good intentions. But we made them because we relied solely on our own ingenuity and our determination to force nature into submission. At first, our decisions seemed like the right thing to do—taking control by substituting artificial wetlands for lost native wetlands. This was hubristic thinking. As we discovered, neither engineering nor determination will get what you want for very long. It took time and failure to admit that we needed to go back to the roots of our training: first, to find out what nature can do for us, and then, to find out what we can do with nature. We knew that native wetlands were far more dependable for sustaining wildlife than anything we could create, but most of the native wetlands had long since disappeared. What did not settle in our minds at first was the idea that native wetlands are created by healthy rivers: if rivers are unhealthy, so are their wetlands. Once this was settled, our mission became clearer.

Rivers play host to literally millions of seasonal migrating waterfowl, songbirds, wading birds, shorebirds, and various other forms of wildlife. The corridors of native rivers had become the last great places for wildlife in West Tennessee. The importance of rivers was mentioned more than two thousand years ago: "And it shall be that every living thing that moves . . . wherever the river flows, shall live" (Ezekiel 47:9). The lesson implicit in this biblical verse applied now more than ever: without rivers there would be few places for many of these living things. Our space-hungry age had left very little space beyond the rivers; now the rivers themselves were nearly consumed.

Reaching around all of this, understanding it ourselves, and explaining it to others were the hardest things we tried to do. As we tried, more than one thousand river miles were reduced to canals, levees, and ditches; miles upon miles of floodplains were left standing in the quiet waters of stagnant swamps. Eighty percent of the natural river wetlands in West Tennessee had been drained, cleared, and converted to other uses, generally from agricultural development. Along with these wetlands went more than a half-million acres of bottomland ecosystems, most of them by the end of the 1970s. Some 350 miles of the native river channels had vanished, enough to make up the entire length of West Tennessee's northernmost river, the Obion.

The public, were it fully aware of the problem, would no doubt take a dim view of the state arbitrarily deciding to eliminate a river. But, in effect, this and much more had been done, and it had happened as much by indecision as by making the wrong decision. While waterfowl managers lost their levees, pumps, and the crops they raised to attract ducks, farmers and the general public lost thousands of acres of cropland, bridges, highways, homes, and other facilities from soil erosion and flood damage. It had gone on for a century. Government help seemed only to make matters worse: without careful analysis and public education, it had been the major sponsor for the heavily modified waterways, and these had failed miserably. The cost has yet to be fully assessed or realized. Most of the tangible costs have been passed on in the form of a perpetual tax burden; the rest have not been fully determined.

And it was not just a local problem. The rivers affected migrating bird life at a continental level. Destroying rivers deprived coastal wetlands and even the oceans of sufficient nutrients and fresh, free-flowing river water. More than 90 percent of freshwater inflow to the Gulf of Mexico originates from the Mississippi River basin of which West Tennessee rivers are a part.[1] There were good reasons to think that these rivers affected even the regional weather.

Rivers are a major function of the regional landscape; nothing stitches the land and its natural resources together into a harmonious tapestry more conspicuously than native rivers. Well-managed, free-flowing rivers preserve the productivity, wildness, and continuity of the land. They are essential for the integrity and character of the country as well as for the livelihood and quality of life for many living things, including us. We saw it all begin to unravel in West Tennessee by the 1980s. Quite possibly, we have not yet seen the worst of it. Suffice it to say that the losses have been enormous, not only for waterfowl and wetland managers, farmers, communities, towns, and cities along these streams, but also for a nation needing clean water and those deprived of the pleasure and enjoyment river wetlands offer. To be sure, the well-being of any river should be a national concern.

In large part, this book records the deeply flawed ways in which the state responded to problems that began some six decades ago. The arbitrary development of wetlands was a tradition generally driven by unexamined preconceptions about how rivers should be managed. The strategy was to impose the human will on the river: if the river refused to behave, then change it. It was deceptively simple but ultimately disastrous. The thinking was one-sided because natural resources like rivers cannot be continually exploited without giving them a chance to recover. Otherwise, the bills eventually come due, and everyone ends up paying for the benefit of the very few. For decades, shortsighted thinking drove the state to employ the U.S. Army Corps of Engineers to ever-greater schemes to control West Tennessee rivers, which seemed to resist such management strategies at every turn. And as the rivers resisted, both government officials and independent land tenants resorted to an even grander form of control—all-out channelization. It was an insidious method of control that, for the promise of short-term profits, involved getting the water off the land as quickly as possible and taking the problem of massive sedimentation with it, no matter how the rivers or the neighbors and facilities downstream would be affected.

To achieve this goal, deep and wide ditches were dug more or less straight, shortening the rivers, gouging them to prescribed depths, increasing their velocity of flow, and sending the sediments elsewhere. This concept was embedded in the thinking of nearly everyone who had anything to do with it and carried out on the largest scale possible. The net result was tragic, as people, plants, and animals that depended upon rivers failed to prosper.

But disappearing wetlands were not just a problem for conservationists and wildlife managers. Politicians and lawyers were also involved, and they clashed, it often seemed, only to practice their trades, frequently doing so without a fair and balanced public input. Personal agendas that disregarded hard evidence and professional advice either dictated how the river would be developed or stymied all progress toward ethical and reliable methods of management—all at a time when the state could ill afford to lose clean, free-flowing rivers or economic prosperity. But the state per se had made no serious attempt to cope with the impending river crisis until farmers during the 1950s and '60s complained about inordinate flooding.

The key "reference rivers" for this book are the Obion and Forked Deer. The Obion is the northwesternmost river in Tennessee. The North Fork of the Obion, the longest of its four major tributaries, flows southwest some fifty to sixty miles until it enters the Mississippi River ten miles or so west of Dyersburg. Reelfoot Lake is a large natural lake and part of the Obion watershed. The Forked Deer River joins the Obion via a manmade canal about three miles before the river's confluence at the Mississippi. Today this complex is called the Obion–Forked Deer River. Since this river complex and Reelfoot Lake were debated, fought over, and studied more than any of the other West Tennessee river wetlands, they are the anchors for the events that unfold throughout this story.

My own sense of the importance of rivers came at an early age. I grew up in the wetlands of northwestern Tennessee, less than five miles from the Mississippi and fifteen miles from the Obion. Until I reached the age of about eighteen, much of my time was spent poling through the cypress swamps and saw grass marshes at Reelfoot Lake. I knew very little about the science of rivers during these early years, but I developed a deep respect for them. My experience was informed by the thought and rumor that the areas along the rivers, heavily forested and sparsely populated, were the wildest places left in this part of Tennessee. Like others living at Reelfoot Lake, I rarely left the immediate environs of this outdoor paradise. The rivers were taken for granted; it was a comfort to think that rivers never changed—that they were eternal.

Using shortcuts, I could walk from the lake to the banks of the Mississippi in about forty-five minutes, although I actually saw that famous river no more than a half-dozen times during my childhood. I did not know it then, but that great river had already been heavily tampered with a hundred years earlier. One thing I did know was that the flooding Mississippi River was important to the vitality of Reelfoot Lake; without it, there would be no source of fish. Also, my duck-hunting friends and I knew that the Obion Bottom and the four rivers south of us were the places where the high-flying mallards went when they ignored our decoys and passed over our duck blinds. So rivers were important, but just how important we underestimated. It never occurred to us that but for the rivers there would be no native wetlands.

I left Tennessee for fourteen years to serve in the military, attend college, and work briefly for the National Wildlife Federation. I returned to the state in 1972 as a wildlife biologist with the Game and Fish Commission and became actively involved in river issues. But the incident that may have triggered much of the thinking that informs this book began earlier, in 1962, as I made my way home after leaving the military. Any change in old and familiar places can quickly grab one's attention after a long absence. But no change was as stunning as what I saw on the Obion River from a bridge on Highway 51. A new, raw, straight ditch went up- and downstream as far as the eye could see. The river I once knew had vanished; piles of fresh dirt were stacked along the riverbank; familiar trees and the bends of the river were missing; and the stream was a sterile-looking chocolate brown. It was a picture of melancholy, even lifelessness.

The "Big Ditch" (left) was part of a huge project by the U.S. Army Corps of Engineers to construct 225 miles of channels throughout the Obion–Forked Deer system. (Photo courtesy of TWRA.) Hydrological chaos in the floodplain (right) was the typical result of such shortsighted thinking. (Photo by Jim Johnson.)

Disturbing as it was, what I witnessed was a classic example of how the state had responded to farmers' concerns about flooding, and indeed it represented the state's main strategy for managing these streams: channelization. That strategy was unfamiliar to most of us in the West Tennessee region before the 1960s; a decade or so later, it had become practically a household word. Key to this awareness was a U.S. Army Corps of Engineers undertaking known as the West Tennessee Tributaries Project (WTTP). The work I had noticed at the Highway 51 bridge was part of this 225-mile channelization project, which was designed to rip through the floodplains of the Obion and Forked Deer rivers. It sprang from a total commitment to the idea that channelization would reduce flooding and promote great prosperity to the farming industry.

My glimpse of the project from the Highway 51 bridge was my first sharp encounter with the deep-seated thinking that drove the state's stream-management strategies. Channelization, the most extreme of these methods, seemed obviously wrong to someone who cherished the pristine and scenic bottomland found only around rivers. Tall pin oaks had once been found throughout these forested bottoms, some so large that two people, joining hands, would have difficulty reaching around them. Red-legged mallards, fresh from the North, spiraled in great numbers in any given year to find nutritious acorns left covered by winter rises in the river. Sport fishing was popular here, especially so in the isolated natural lakes known as oxbows. It was common to hear that commercial fishermen had caught flathead catfish weighing fifty to sixty pounds in the deep holes found in river eddies. These images formed my vision of the river.

But then the vision changed. So rivers were not forever, as I erroneously believed at an earlier time. Had we invented anti-rivers? I began to think so after I became the supervisor of the state's wildlife lands for the northwest region of the state. All of our wildlife areas were along the rivers, and nearly every acre was within the floodplains of rivers. Three main wildlife management areas (WMAs) would play a pivotal role in this story—Reelfoot Lake WMA, Gooch WMA, and Tigrett WMA—and all of them deteriorated because their rivers had practically ceased to function as major native waterways.

Our earliest encounters with conflicts between modified landscapes and native wetlands came at Reelfoot, the state's largest natural lake. It was in dealing with the consequences of landscape changes in this WMA during the mid-1980s that we learned many of the basic principles that finally coaxed us to use better methods for managing wetlands—that is, to be guided by the natural principles that caused the wetlands to survive and flourish, whether native or artificial. Since Reelfoot was once a wetland nurtured and produced by a native river—the Mississippi—the principles learned from her aptly applied to the West Tennessee tributaries.

Gooch WMA on the Obion was an artificial wetland created in a natural setting, a "model" for waterfowl hunting areas during the 1960s. Twenty years later, it was considered a grave mistake because it had failed to meet the state's waterfowl objectives. It was also costly to maintain and contributed to the destruction of the river. Changing the management of this waterfowl area became a political nightmare; it forced us use methods we knew were ethically wrong and would not work.

Tigrett WMA was on the Forked Deer, and a central part of this story is the twenty-five-year effort by wetland managers to restore the Tigrett floodplains. The plan met strong opposition from political interests tied to the U.S. Army Corps of Engineers' WTTP, an ill-conceived channelization project. The challenges were compounded when conservationists who ostensibly could have saved the rivers from certain disaster became part of the problem.

Could the rivers of West Tennessee ever be fixed? The advice offered upon my leaving the National Wildlife Federation to work for Tennessee was not encouraging. "You don't want to go there," William "Bill" Reavley, my boss and field director, told me. "You can't do anything in state government." But I had to think it would be different in Tennessee. The next question was how it all should begin. First, our agency's policies for managing wetlands had to change, and the state had to change its policy for managing rivers. How could these changes be kept on track? My colleagues and I would fly over the rivers many times wondering if any of our work made a thimbleful of difference. The struggle is ongoing and remains paramount throughout this saga of West Tennessee rivers.

Part 1

The Pathway to Disaster

Chapter 1

The Five Rivers along the Mississippi

In the lowlands of West Tennessee, five major tributaries flow into the lower Mississippi River: the Obion, the Forked Deer, the Hatchie, the Loosahatchie, and the Wolf. Draining from the soft topsoils of a terrain that ranges from hilly to nearly flat, these rivers are unlike such sister streams as the Tennessee and Cumberland rivers, which drain (via the Ohio) into the upper Mississippi from the steep, rocky terrain of East Tennessee. The five westernmost rivers have proven particularly vulnerable to human interventions and, for the most part, are now only shadows of their former selves.

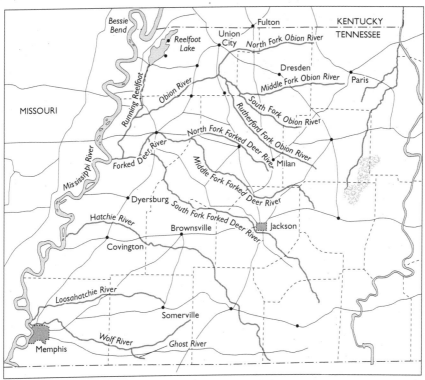

The rivers of West Tennessee, 1938.

The soils in the watersheds of the five rivers in West Tennessee are uniquely different from those of the rivers in the east. The soils of the eastern region have relatively hard surfaces, while those of the western region are mostly soft and windblown. The upper horizons of the soils in West Tennessee are loess soils (a mixture of sand, clay, and silt chiefly deposited by winds) layered over sandy soils. They are highly subject to erosion. Although the five rivers and their watersheds have individual characteristics, the effects of soil erosion and the subsequent deposition of sand in the streams are very similar, as are the human methods used to correct these problems and the ultimate consequences of those interventions. In each case, the heavy footsteps of civilization have set up these rivers for failure: once the watershed is disturbed beyond a certain point, the fragile topsoils are washed away. Eventually, even the sand horizon beneath the topsoil is laid bare, and it too is washed into the streams. Sand is relatively heavy and resists being transported by a stream. It accumulates and the debris that lodges in these accumulations exacerbates the build-up. The streams thus fail to function at their former capacity. The next thing that occurs is abnormal flooding, and the human populations that are affected by such inundations demand, of course, that something be done.

Channelization, a simplistic solution, addresses the problem of flooding directly. It is a single-minded method that shows very little, if any, consideration for its future consequences. The method of choice for societies on the move, it is also a quick fix that does not address the root cause of the problem and largely ignores the dynamics of the former stream. This is what has happened in West Tennessee as most of the native streams were replaced by human-built channels. Channelization has practically sealed the fate of these rivers: natural hydrology could not be sustained, and the rivers have been set on a course to ruin. Nothing can change this ultimate destiny until the rivers are relieved of the stress placed upon them and their natural hydrology is reestablished.

Every river in West Tennessee shows the impact of these encroachments—including the Hatchie. But while the Hatchie River has not escaped the debilitating effects of soil erosion, it has, at least, escaped extensive channelization to its main corridor. This is an enormous break for the river as well as for Tennesseans. Now the Hatchie survives, albeit tenuously, as a local model for those, like wetland managers, who have recognized past mistakes.

Wetland and other natural resource managers often use a scale of one to ten in their quick field assessments of the relative health and vigor of a river and sometimes other resources, such as plant and animal populations. The rating of one, for example, means that the relative health of the subject or population is extremely poor, next to nonexistent. A river with this rating has lost most of its natural stream characteristics, whereas a river with a ten rating represents the best that a river can be—that is, one unaltered by humans. Using this scale, the Mississippi and its five tributaries in West Tennessee would all have been rated ten before human intervention. Today the Hatchie River, with a rating of seven, remains our best example in the region of what is needed for the others, which are rated five and below. Only recently have we had the will and determination to understand and cope with the failure of these rivers.

Rivers need not be our nemeses; they can be our allies. While no one expects that making this happen will be easy, the potential is enormous. Although the task may seem thankless now, there is little doubt that whoever is able to see it through successfully will reap the gratitude and appreciation of future generations. To be sure, the will and the commitment are all that remain to realize the goal of revitalizing West Tennessee's rivers. The technology is available, and there is even a plan in place to effect the recovery.

The following descriptions of the rivers along the Mississippi begin in the south with the Wolf at the Tennessee-Mississippi state line and end in the north with the Obion at the Tennessee-Kentucky state line. Generally, the impact from channelization is more pronounced as we go from south to north. This trend mainly reflects the characteristics of topography: the lower-gradient land (i.e., the better farmland) lies to the north while the land to the south becomes increasingly hilly.

The Wolf River

From its confluence with the Mississippi through approximately fifteen miles of metropolitan Memphis, the Wolf is an artificial river. Then, upstream for thirty miles or more, the main corridor of the river proceeds without evidence of channelization; yet, nearly all of its tributaries show signs of this work. Upstream from Germantown, the river is scenic since it still meanders and trees survive along most of the floodplain. Segments of the river remain popular for boating, fishing, and canoeing, especially in the area between the towns of LaGrange and Moscow.

Conservationists have campaigned for the protection of the Wolf, particularly in the segment between Early Grove and Tena creeks at Rossville. Part of this area is known as Ghost River. The name is probably derived from the swampy haunts that extend three or four miles upstream from the "valley plug" that created it. Evidence indicates that a heavy deposit of sand from an adjacent creek created the valley plug, or dam, backing up the Ghost River reach. Young cypress trees grow thick in one section, but water tupelos are dead or dying throughout the lower part of the swamp.

Some conservationists have lauded the Ghost River area as an extraordinary natural gift, the beneficial byproduct of natural ponding. However, the sediment plug that created the area was no doubt caused by human activity. These conservationists have not yet realized that the trees (even the hardy cypress) cannot live in permanent standing water; sooner or later they will die prematurely.

Cypress trees do not sprout in standing water; like all trees, they require dry land in order to sprout. After the trees start growing, the roots must undergo seasonal drying, at least every few years, to live in a healthy condition. Occasionally, there are exceptions: cypress trees live to a ripe old age in certain southern perennial swamps, such as the Okefenokee on the Georgia-Florida border. Cypress trees require oxygen, absorbed through their roots, to survive, and while stagnant swamps do not contain enough oxygen to sustain them, the constant flow found in perennial swamps is sufficient to remove toxic substances and gases that accumulate under still-water conditions and to oxygenate the water and provide the trees with what they need.[1] History shows that these swamps were once dry—a necessary condition for all trees to sprout from seed and become established.

The support among some conservationists for artificial swamps reflects a lack of understanding of these natural processes. The Tennessee Wildlife Resources Agency (TWRA) recently established the Wolf River Wildlife Management Area at the Ghost River segment, but the agency is likely to discover that, without prudent management, the entire segment may become as decadent as the artificial swamps at Tigrett WMA, which will be discussed at length in the pages ahead.

The Wolf is in an advanced stage of stress throughout its length. Like its four sister rivers to the north, the stream suffers from tributary channelization. This channel has filled with sand, causing premature overflow and prolonged or permanent water on the floodplain. The upper reaches of the main channel break down into several smaller streams, a process known as braiding. Viewed from the air, odd "bulges" appear along the length of the river; it looks like a sick rat snake trying to digest indigestible objects.

Braiding, a natural adaptation by the river, occurs when obstruction of the main channel (usually started by accumulations of sand) becomes excessive and the energy of the stream is unable to move it. Such obstructions prevent the stream from carrying its normal volume of water. To compensate for the lost capacity, the stream reestablishes itself by separating, often into many smaller channels, and the process continues until the capacity is met.

This self-maintenance is common in streams where channelization does not interfere. The solution is to stop the erosion that produces the sediments. Since the main stem of the Wolf River has not been channeled, it continues to self-adapt to survive the abuse. Although its floodplain is stressed and becomes wetter, the area along the river still displays characteristics of natural wetlands, in fact showing an extraordinary tenacity and capacity for survival. Rating about a five on the one-to-ten scale, the Wolf is a good candidate for stream restoration.

The Loosahatchie River

The Loosahatchie is the shortest river in West Tennessee and was once a fork of the Wolf. Like the Forked Deer, however, it was modified prior to 1913 by civilization and is now an independent river.[2] The Loosahatchie now enters the Mississippi River approximately two miles upstream from the Wolf River within the suburbs of Memphis. Excavated ditches characterize the Loosahatchie; it has no natural tributaries.

This river functions essentially as a ditch to expedite runoff from farmland. It probably transports as much debris from the Memphis suburbs as it does sediments. For this reason it is correctly identified as the Loosahatchie Canal. There are a few forested sections with artificial pools along the stream but not much more. Some creativity will be necessary to reestablish the river to a seminatural state. The Loosahatchie rates about a two on the one-to-ten scale.

The Hatchie River

The Hatchie, with its aforementioned seven rating, is considered the only natural river in West Tennessee, but how natural is subject to debate. Most of its tributaries have been channeled by private developers or through the help of the Soil Conservation Service (SCS) and its former Public Law 566 projects. There was considerable public speculation in the early 1960s that the Corps of Engineers had plans to channelize the Hatchie. State legislation, however, thwarted this plan by classifying the stream in 1968 as a State Scenic River. For once the public took a stand against the destruction of a West Tennessee stream, and thus it was spared from main-stem channelization.

The Hatchie River, meandering and shifting in the floodplain, has created wetlands in West Tennessee that remain much more vigorous than those of its sister streams. (Photo by Jim Johnson.)

Farming and other developments on the Hatchie usually stop at the edge of the river valley. Nevertheless, encroachment on the floodplain has occurred, and there is always the threat of further encroachment. Even so, the Hatchie remains the best example in West Tennessee of a relatively natural river. The public is prouder of it every year, and so are the wetland managers and conservationists such as the Nature Conservancy.

The Hatchie enters the Mississippi twenty-five miles above Memphis at Fort Pillow State Park and wiggles for thirty-five miles upstream. It has no major forks, only creeks and branches. Its wide valley is filled with bottomland hardwoods, eighty to ninety feet in height, for eighty miles or more beyond the town of Bolivar.

Although the Hatchie is the best river we have, its upper part, from Bolivar to its headwaters, is far from ideal; this segment is in complete hydrological disorder. Dead and dying trees can be seen from every road that lies near this reach of the river. If a single tributary on this river has not been channelized, I am unaware of it. At the upper extremes of the Hatchie is one of the largest tributaries, Porter's Creek. Here, the repeated ditching and the construction of levees to protect cropland are much like the interventions in the floodplains of the Obion–Forked Deer. The evidence of soil erosion and small channelization projects found here recalls precisely how the deterioration of the Obion and Forked Deer began.

From a few miles below Bolivar and upstream to the Mississippi state line, the river braids in pulses similar to the upstream reaches of the Wolf River; these indicate that the Hatchie, like the Wolf, has foundered on sand. Valley plugs and artificial swamps have resulted in some locations and hundreds of acres of hardwoods have died because of the swamps.

For more than twenty years I observed the river as a member of the Hatchie-Rice Hunting Club, whose hunting area bordered the river for two miles across from the Hatchie National Wildlife Refuge. This section contains some of the best timber along the Hatchie. During this time, the floodwaters have subtly but surely increased in duration on its floodplain. Some of the tall hickories (trees that I once depended on for a limit of fox squirrels) are stressed more each year from the prolonged flooding. Other tree species have also suffered, while more water-tolerant trees like swamp white oaks, sweet gum, and river birch have replaced the original forest at many sites. The river channel has become shallower from sand deposits and wider than normal as it tries to compensate for the accumulated sand and the channel's lost flowage capacity. The oxbows also fill at a rapid rate, a sure indicator that an inordinate amount of the heavy sediment is entering the floodplain. Deep holes and nooks created by channel variations in the main channel are needed for sport fish such as catfish, bluegill, and large mouth bass, but these are limited or missing.

Despite this, the Hatchie is ecologically the healthiest river in the western part of the state. It still has approximately 140,000 acres of bottomland hardwoods, estimated to be more than its four sister rivers combined, and more than any river in Tennessee. Various statistical comparisons that have been made between the Obion–Forked Deer complex and the Hatchie offer further evidence of the latter ecosystem's relative health. For example, data collected on timber kills from artificial swamps on the Forked Deer showed that 23.4 percent of the timber (124,755 acres) was dead or dying between the 1950s and the 1980s. By contrast, on the Hatchie during the same period, only 3.2 percent of the timber (5,434 acres) was similarly affected. Studies also showed that the Hatchie had far less sediment entering its waters. Whereas sediments measuring some 25 tons per acre eroded into the Forked Deer floodplain, the sediment accumulation in the Hatchie measured only 7.1 tons per acre.[3] The numbers for aquatic bottom-dwellers, biomass (overall amount of living matter), and species diversity, particularly that of birds, have also been higher for the Hatchie.[4]

Even though the Hatchie is known for its upland game—such as whitetail deer, turkey, squirrels, and swamp rabbits—it can also host good duck hunting. Spence Dupree, who lives in Jackson and was born and raised on the Hatchie, describes some of the early days of such duck hunting in his book, *Across the River:*

> Fall and the coming winter meant but one thing to us, duck hunting. We literally prayed for rain so the Hatchie would spill over its banks. The swollen river and its feeder creeks would burst out over the floodplain covering big pinoak flats to form prime wintering grounds for waterfowl. Then the mallards would come. The trick was to find them and stay with them as water levels changed.

A winding tributary of the Hatchie River, still a healthy and aesthetically pleasing stream. (Photo by Jim Johnson.)

The Five Rivers along the Mississippi

It is somewhat an art form to keep up with the concentrations of mallards when water levels are varying continuously. As the water rose in the Hatchie, ducks generally moved toward the headwaters. Conversely, as it fell they followed the crest as it worked its way toward the Mississippi.

But even the Hatchie had some fields to hunt. "This is particularly true of any functioning river system in agricultural land," Dupree writes, adding,

Ducks [also] feed on crops like soybeans. These fields flood on high water and waterfowl will normally follow the crest.

This was duck hunting in its finest form. There is no blind. There is no leased land near a road, or warm clubhouse to return to if socks get wet or guns freeze. The only boundary is the water's edge. This is what Nash Buckingham meant when he spoke of "rough and tumble ducking."

Dupree also notes that when the Hatchie drains, it is good for upland game.[5]

The Obion–Forked Deer Complex

This river complex and the issues surrounding it are the primary focus of this book. No state should allow its rivers to fall into the condition of the Obion and the Forked Deer, which were individual streams until the state government and its partner, the U.S. Army Corps of Engineers, decided to modify them. Now the two rivers join for the last three miles of the main channel before entering the Mississippi River. The corps began its enormous channelization project in 1961, initiating a longstanding controversy. Ten years later, the legal battles over this work were only beginning to brew. Today, they still brew.

Economic growth in a fourteen-county area in northwestern Tennessee (the watershed of the Obion and Forked Deer rivers) has lagged behind the rest of the state. The inability of these rivers to drain sufficiently is considered to be a major cause of the problem since cropland was affected and agriculture was the major industry in this region. According to a 1972 report by the U.S. Department of Agriculture, the floodplains of these rivers encompassed 759,000 acres, 3 percent of the state. These floodplains were said to produce more than 30 percent of the state's soybeans and also accounted for 35 percent of the state's total bottomland hardwood production.[6]

According to the same USDA report, crop damage from flooding was estimated to be $9.8 million annually, with damage to roads and bridges about the same—$9.7 million. With the costs of other damage (including flood damage to residential, commercial, and industrial property, as well as indirect and sediment damage to crops and forests) added in, the grand total came to $31 million. This was a hefty cost, especially considering that the expected annual profit from crop production in the 140,000 acres affected was $32 per acre, or only about $4.5 million.[7] What economic sense did it make to continue practices that created more damage than benefits? It seemed better to give the floodplain back to the river.

While the economic questions concerning the impact of flooding on agriculture were always kept in the forefront, other possible economic issues were ignored. Officials and waterway managers might have asked how much income could have been generated from duck hunting, fishing, good drainage, and the harvest of hardwood timber. Or they might have asked how much damage could have been prevented to cropland, bridges, roads, and other infrastructures within the floodplain had the proper criteria been used for the development and management of the floodplain. With regard to these questions, no one really knew what the economic costs of channelization were. Comparing the costs and benefits of maintaining the Hatchie River, where channelization was limited, to the economic picture at the Obion–Forked Deer could have yielded some very interesting data. While common sense suggests that the benefits within the Hatchie floodplains have exceeded the costs several times over, no comparative analysis has ever been done.

The wetland complex in these river basins provides a winter home for more than half of the waterfowl in the state. Over one-third of the floodplains, by the corps's estimate, was considered "prime" habitat for migratory birds, including highly valued ducks and resident wildlife. But prime habitat depends on who identifies it. Waterfowl managers in the northwest district do not agree that dead trees and stagnant swamps are prime waterfowl habitat, but by my reckoning, such degraded acreage made up about 80 percent of the corps's estimate.

A closer look at the two main parts of the river system will help us better understand the problems afflicting these floodplains.

The Forked Deer River

The Forked Deer River has three major branches—the North Fork, the Middle Fork, and the South Fork. The Middle Fork, or the main channel, is the longest of its three major branches, and today it flows along stretches of straight channel for more than seventy miles. Until the mid-1820s, the Forked Deer went an additional fifteen to twenty miles farther south, but that changed with the first efforts to modify the river. In 1823 a canal was dug in Dyer County (established that same year) from the Forked Deer channel to the Obion River, entering the Obion about three miles from the Mississippi River and cutting off the lower portion of the Forked Deer. This project, like all the others that have modified the rivers and wetlands in West Tennessee, likely sprang from impulsive thinking, not common-sense thinking. During the early nineteenth century, as in later eras, the impulse was to conquer and control all things natural. Such thinking has long been considered "progressive."

All it required in the 1820s to justify river channelization was a simple logjam caused by the local timber business. A 1995 article, written by local historian Earl Willoughby and published in the *Dyersburg State Gazette,* detailed the first canal's origins: "The lower reach of the Forked Deer was so jammed with abandoned boats that navigation was beginning to become very hazardous." As Willoughby noted, Colonel Robert H. Dyer (the landowner and timber dealer for whom the county and the town of Dyersburg were both named) applied to the state for funds to build the canal, which would provide a shorter and safer route to the Mississippi. "Between the two rivers lay a low, swampy area and in times of high water, the over flow would connect the two rivers with Reelfoot Lake," Willoughby wrote. "The region was often referred to as Wood Lake. Col. Dyer won the funding for his venture and began the canal—which conveniently ran past his warehouse."[8]

As it stands today, the lower portion of the Forked Deer below Moss Island WMA, once the most dynamic segment of the river, lies withering away in Lauderdale County, practically filled with sediments. Without a source of water, this thirty-mile section will disappear. Controversy still exists over this canal, which has no significant purpose, and it appears that every effort is being made to level and drain the old, abandoned, lower part

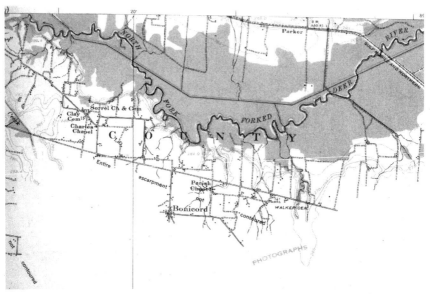

North Fork of Forked Deer River in 1939, showing relatively natural hydrology.

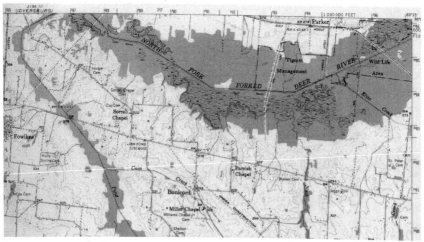

North Fork of Forked Deer River in 1961, showing diversion of the waterways and poor hydrology due to excessive channelization.

The Five Rivers along the Mississippi

of the river for cropland. Officials continue to brood over strategies to restore this lost segment of river. But while natural resource managers await the outcome, time passes and more reasons are found not to bother with restoration. Before long, the old river scar will likely be filled and farmed, and the river—and perhaps its history—will be lost.

It becomes increasingly difficult to find any good effects from channelization. The Forked Deer River, including its Nixon Creek branch in Haywood County, has lost at least seventy-three miles of its original channels. The extensive misuse and abuse are nearly as great on the Forked Deer as on the Loosahatchie. Hardly a tributary has avoided channelization. Channelization on the river's main thoroughfare occurs from its confluence at the Obion River to its headwaters. In effect, the entire river is left in ruins, and the landowners and other users of this natural resource feel the effects.

From the air, the scattered dead and dying swamps appear as pockmarks on the floodplain—though, to some misguided souls, they may appear to be sparkling oases. Segments of the old meandering river lie alternately along the channel like the hacked pieces of a despised snake. Large artificial swamps appear randomly throughout the Tigrett Wildlife Management Area, just as they do throughout most of the upper reaches of the river. The once abundant trees are gone. Nothing about the river appears to be in a thriving state, except for a few tributaries in the upper reaches, and a few patches of isolated trees here and there.

Access to the main channel is limited or inaccessible because of the impenetrable swamps. The swamps are practically inaccessible except at highway crossings, the ends of abandoned roads, and a few boat ramps. Upland game and other wildlife can be found in only a few places, mainly because of the limited access and the lack of dry ground. Fishing holes are rare, found only in a few meander scars or small oxbow lakes. Most of the swamps are too depleted of oxygen to sustain fish.

This lateral ditch on the Forked Deer River, which does not follow the natural drain in the floodplain, has contributed to the failure of natural hydrology and the destruction of the river. (Photo courtesy of TWRA.)

Despite these devastating environmental problems, waterfowl hunting is remarkably good in a few isolated places where habitats occur at the edge of the floodplain or just outside it. Horns Bluff Refuge, midway on the Middle Fork, for example, hosts some ten to twenty thousand ducks through the winter and provides good waterfowl hunting adjacent to its north boundary. Overall, however, prime waterfowl habitats have declined significantly throughout the Forked Deer River basin. This decline will continue without the intervention of management.

The recently developed South Fork WMA lies in the floodplain near Jackson. This area is in an early stage of development, restricted by politics and by its status as a mitigation land bank for the Tennessee Department of Transportation. Although this WMA provides good waterfowl hunting, its future use has not been fully determined.

Other forms of outdoor recreation—hiking, boating, birding—have not fared well at all. Indeed, owing to the lack of access and swamped-out floodplains, such uses are virtually absent from 90 percent or more of the river.

The situation is far from hopeless, however. While the Forked Deer River rates only a three on the one-to-ten scale, it is an excellent candidate for restoration.

The Obion River

The Obion River (called the Bayou River in the early 1800s) flows southwest from its major branches, the North Fork, Middle Fork, South Fork, and the Rutherford Fork. Rating a four on the river-condition scale, the river ends at mile 821.4 on the Mississippi River. Like the Forked Deer River, the Obion is historically famous for duck hunting. Waterfowl hunting in the Obion River bottoms is described in nearly every article and book written on the subject in West Tennessee. The area is well known even today for this activity. As has happened with other channelized rivers, cropland, artificial swamps, or other developments have displaced natural waterfowl land that once attracted ducks during seasonal floods. Consequently, the methods of hunting have had to change with the landscape. As on the Forked Deer River, the heyday of duck hunting in a natural environment is over. Gone are the days of hunting the tremendous expanses of flooded oaks during the rainy periods of fall and winter. Agricultural encroachment has drained traditional wintering grounds, and flooding has killed the oak trees and other favorite sources of waterfowl foods.

Nonetheless, famous private waterfowl-hunting organizations, like the Davy Crockett Hunting Club, today are still well known on the Obion. The Grissom Farms, the J. Clark Akers Farm, and other places upstream and downstream also provide good hunting. But while hunting is generally good at these places, site preparation and maintenance remain costly. The grounds of most of the hunting clubs are on the riverbanks where annual flooding is persistent and damages or destroys the developments. Nearly all of the hunting developments depend on river water, costly pumps, culverts, and levees for support. Most of these facilities and structures have a lifespan of ten to fifteen years, provided that the river does not take them. The gouging force of the river is a relentless and constant menace to the structures, and thus they must be replaced on a regular basis. Managing waterfowl developments along the river costs as much as ten times more than what it costs on better-located sites above the annual floodplain. Today, TWRA maintains several waterfowl units along the Obion. They suffer the same problems as the private developments.

Like a string of oases, the WMAs start at Moss Island WMA on the lower Obion River and continue to the Jarrell Switch wetland between the towns of Trezevant and McKenzie, an area that lies near the headwaters of the South Fork Obion River. At three-mile intervals up the Obion from Moss Island, Ernest Rice, Sr., WMA, White Lake Refuge, and Bogota WMA lie in succession. Thirty miles farther up the river is Gooch WMA, located at U.S. Highway 51. The TWRA's hope is, eventually, to fill in the gap between Bogota and Gooch in its quest to salvage wetlands and duck hunting. A succession of WMAs continues upstream from Gooch for another forty miles.

The total acreage under the WMA program on the Obion currently exceeds thirty thousand acres. Another twenty thousand acres are expected to be added within the near future. Nearly all of this land will likely be in the river floodplain, unless administrators take heed of the experiences and advice of wetland managers. The TWRA faces a strong challenge in managing these lands while minimizing damage and enhancing the natural functions of the river. If nothing else, public ownership offers an opportunity to manage the river in an ethical and equitable way.

Chapter 2

River Users

Ultimately, rivers are for people. How we use them determines how well and how much they benefit us. Whatever affection some of us might have for rivers, they have no respect for us or our civilization. Unfortunately, since the advent of the machine age, this business of "no respect" has become mutual, as we have used our technology to force rivers to our will. However, disrespecting river resources has proven unnecessary and ultimately unprofitable.

On the other hand, respecting rivers does not mean that they should be left exclusively pristine and untouched. All of their native characteristics and functions are renewable and sustainable with good land stewardship. The rivers are natural resource banks with many savings accounts: timber, wildlife, aesthetics, and much more. Once the surplus in any account—its accumulated interest, so to speak—is withdrawn, the accounts must be allowed to regenerate. The surplus can be taken out at one time or withdrawn over the course of a year, but the stock or principal cannot be touched without weakening the account. If too many accounts are overdrawn, the bank—that is, the functional river and its resources—becomes bankrupt. Nothing but a shell may survive. In the case of the Obion, for example, little is left of the river but the valley itself, and much of that has ended up like an abandoned bank—wrecked and ruined. What is surprising is that we have had no warnings from reputable economists. Preserving rivers for equitable and sustained use requires conscientious land-use planning and management. It also requires compliance with well-conceived rules designed to serve the many river users in the long term, and this must be accomplished by the strong but fair arm of the public. But this has not been done.

A river has innumerable users. Located near cities and towns like Memphis, Jackson, Covington, Union City, Dyersburg, and Martin, the rivers of West Tennessee have more human users than most, particularly since they run through lands that are heavily agricultural. Some of the most concerned users of these river ecosystems have been sportsmen—people who hunt and fish—often because there are so few places in the area for pristine outdoor recreation. While the struggle to preserve river ecosystems has often been waged against those seeking short-term economic gain, not all who push for development in the watersheds and floodplains are mindless enough to wreck the rivers, especially at the expense of others. Developers, farmers, and landowners can also be river advocates and frequently are.

There is little doubt that the public is sensitive about what happens to rivers when they understand the problems. The difficulty is that, too often, they have no organized methods or means to collectively focus this interest. Those who abuse the rivers are pleased that this interest does not surface as a single voice. But the people of the state must assert themselves if their rightful use of a river is to be protected and realized. To restore, say, one hundred miles of the Obion River within ten years, the people need to take one determined step after another, keeping their focus on the ultimate purpose. They must find and implement the best methods that will eventually restore the entire river for the general good. Railroad companies and public highway departments, for example, must be convinced that they should build tracks and roads across the floodway with open-span bridges. The problems should not be discouraging but set aside and dealt with individually. As painful as the steps along the way may be, these problems can eventually be resolved. There are, in fact, many examples where such problems have been dealt with successfully, sometimes in local courts when there was no other recourse. Considering the inevitable, rising cost of damages and the lost benefits if the rivers continue in their present state, the steps must be continued for as long as it takes to accomplish the goals, whatever the obstacles.

In West Tennessee there have been no large-scale referendums, polls, investigations, or other means to determine what the public thinks about rivers, what they need, or what they are willing to support. Most of the rivers in West Tennessee have long since lost their popularity for outdoor recreation. Outdoor use on most of these streams has dropped to such a level that there are now few people around with enough familiarity of the rivers to comment on their status. The most common users in the past were hunters and fishermen, but their usage is now lower than it has ever been. By the mid-1980s, even TWRA fisheries scientists had largely stopped collecting data on fish populations and fishing for all the rivers but the Hatchie. Most of their stream surveys are now limited to small, relatively undisturbed tributaries in the upper watersheds. Consequently, information about the use of West Tennessee tributaries comes largely from wetland managers or county wildlife officers who make observations incidental to their routine assignments.

It has been left to those using the rivers—landowners, a handful of sportsmen, and those with political agendas—to complain. The general public, with no organized means to express their opinions, have remained the silent majority. State conservation and natural resource departments and private conservation organizations have made no concerted or joint effort to inform them. Nor has there been a noteworthy national effort. The media have done their best, but the exercise has been as daunting to them as it has been to wildlife officers. No one has been able to grasp the magnitude of what was actually lost or what stood to be gained by restoring the rivers. This was particularly true in the 1970s and '80s, and while hopeful signs had surfaced by the 1990s, the promise of restoring the rivers remains mostly unrealized.

Better insight into what we have lost and what we still stand to gain can be gleaned from a closer look at the key groups of people who use the rivers.

The Game Hunters

While no one has used significant stretches of the rivers, particularly the Obion and Forked Deer, more than waterfowl hunters, they are hardly the only sportsmen who frequent the riverine areas of West Tennessee. Along the Hatchie River and some reaches of the Wolf, other kinds of hunters have stalked their own favorite game—including squirrels, whitetail deer, wild turkeys, and raccoons. The abundance of such game in those areas has been largely due to the surviving bottomland hardwoods.

Together, the sportsmen using the rivers—known as the "River Rats" in the regional vernacular—have been acutely aware that maintaining the health of the forests and wetlands is all-important. Their compassion for the rivers is year-round, and thus they have been prime candidates to join forces against anyone who threatened those cherished acres.

The Hatchie "River Rats" are not naïve about the purpose of "their" bottomlands; they know that the trees there survive mainly because timber is more profitable than cleared land. They doubtless know that the land's value to timber companies is the reason why the Hatchie is still free to flow. While they might become melancholy when the trees are harvested, they try to be philosophical about it. "It'll make good deer habitat," they might say. Or, "You know they cut the old Ashby Place about twenty years ago, and that place looks almost like it was never cut. It works pretty well if they don't cut the whole tract in the same cutting." The hunter owes considerable gratitude to landowners practicing responsible land stewardship, as has occurred on the Powell, Ashby, and Alexander properties, all huge tracts of timberland along the Hatchie.

No one locally ever refers to these large tracts as "forests." Anyone using that word, as appropriate as it might be, is considered a stranger. The locals call these tracts "the woods" when they are small, or "the bottoms" or "bottomland timber" when they are large. Such terms describe perfectly how "River Rats" feel about this land. The words "the bottoms" ring in their minds like the echo of a falling hickory nut though a canopy of tall, fully mature hardwoods. The nut's solid "thump" on the forest floor leaves one with the gratified feeling that the place is complete.

Even with the poor condition of the Obion and Forked Deer rivers, many sportsmen—mostly local hunters—still find a few isolated forested tracts on these streams for the treasured experience of hunting deer, wild turkeys, raccoons, and squirrels. Such hunting, however, is largely limited to the edge of the floodplain at field borders or where live timber still grows. but it is not guaranteed. Ask hunters, and most will tell you that waterfowl are the big game, the small game—indeed, the all-game—along the Obion and Forked Deer.

In 2003 waterfowl hunters were estimated to be about 28,000 for the entire state. But this number fluctuates with good and bad seasons.[1] An estimated 75 percent of these hunters prefer their hunting within the Obion–Forked Deer system; indeed, hunters in northwest Tennessee have accounted for about 50 percent of the state's total duck harvests, which average about 275,400 birds per year.[2] The ducks have been coming to these river bottoms for centuries, which make them an important wintering ground not only for the state but also for the Mississippi Flyway as a whole. Extrapolations from a midwinter survey in 2002 indicates that Tennessee harbored about 6.5 percent of the ducks inventoried in the states along the Mississippi Flyway.[3]

The Mississippi Flyway is said to be the richest migratory bird corridor in the world. Frederick Lincoln noted in 1935 that anyone "stationed at favorable points in the Mississippi Valley during the height of migration can see a greater number of species and individuals than can be noted anywhere else in the world."[4] This probably still applies. The Mississippi Flyway has more wetlands and more ducks than any flyway in the United States. It contains more than half the acreage of the country's wetlands; it generates sales of 40 percent of the duck stamps that hunters are required to purchase; and it is where hunters kill nearly 40 percent of the total ducks harvested in the United States.[5]

The way hunters feel about rivers matters greatly in how they will be managed. In West Tennessee some are still content with their isolated hunting areas, but others remember what the hunting was like when the rivers flowed reasonably well and oaks filled the river bottoms. Hunters in the Obion and Forked Deer bottoms have usually managed to get their share of ducks in the flyway.[6] Hundreds of hunting sites are established along these rivers. From the rivers' headwaters to their confluence at the Mississippi, nearly every nook along the streams during duck season has a blind with hopeful hunters. These avid waterfowlers make it a year-round ritual to either search for a place to hunt or to spruce up an existing site.

Spence Dupree, who lives in Jackson, is one who knows the passion of duck hunting in both the Hatchie and the Obion–Forked Deer bottoms. He is certainly not alone, but he has a special talent for describing the "the way it was" for hunts in both of these venues. Dupree hunted Twin Rivers, a club located in the Forked Deer River bottoms only a short distance upstream from the town of Halls. Unlike duck hunting on the Hatchie, most of the hunting on the Obion–Forked Deer was usually from the comfort of a duck blind—not one made from sprigs of cane and brush as in the old days but a camouflaged box with comforts closer to that of the den back at the house.

Duck blinds have always been an issue to reckon with when managing wetlands. There are many duck blinds in places city folk might call "wasteland." But for those who long to hear the friction of wind howling to a fine pitch through the primary feathers of cupped wings, a duck blind in these wastelands can be more famous than—and outlast generations of—the renowned duck hunters laying claim to it. "A duck blind is a place where the best and the worst of men may readily show," Dupree wrote. "That was proven many times at Twin Rivers. From my viewpoint, when it came to our bunch, it was always the best."[7]

When Dupree and eleven others in the club would meet to discuss the upcoming season, part of the discussion involved the blinds: Were they in the right location? Were they camouflaged properly? And what about the pumps and water-control structures: Did they work? If not, how much would they cost to fix? The list went on. The talk of "fixes" could be heard in almost any duck-hunting club in these river bottoms.

In particular, fixes to the levees, pumps, and water-control structures could be the club's nemesis. There would always be a difference of opinion about whether these structures were doing what they are supposed to do. Not every club was as fortunate as the Twin Rivers club. In some clubs, a little knowledge could be a dangerous thing, and when the wrong person was given sole responsibility for water control, foul-ups in the water-level regime in the hunting fields could occur. "But it seemed Providence meant for us to have the duck club," Dupree observed. "Several members had special talents necessary to make it work."[8]

The planning sessions at Twin Rivers and other clubs were just as important to the hunt as the hunt itself. It made the members comrades. They needed to be since they spent countless hours crammed together in those boxes called duck blinds. Such blinds usually measure from six feet to eight feet wide and fifteen to twenty feet long. They could house eight to ten hunters, along with guns, shells, extra coats, lunch sacks, a butane heater or two, and a restless retriever. Some are even much larger, with cooking stoves, refrigerators, and television sets, but this was not the case at Twin Rivers. That club's duck blinds could be pretty crowded camping quarters on the weekends and holidays when the young men—duck hunting is an almost completely male sport—came to hunt.

"As opening day neared," Dupree recounted, "the bottoms were dry with little or no sheet water in the fields other than those that pumped. While we had water, it covered

Veteran duck hunters at Reelfoot Lake take a break to swap stories in the kitchen of the Barker-Fraley Duck Blind.
(Photo by Jim Johnson.)

only 30 percent of what should have been the normal area. Still it was with great expectations we took to the blinds on the first hunting day of the Twin Rivers Duck Club. Our optimism was in vain. The three blinds killed a total of one mallard."[9]

But the days of low productivity from the hunt were soon forgotten on the Forked Deer. If only one day of good hunting came along, it was enough to be optimistic. "This time it rained ducks," Dupree wrote of one day that mattered. "I count eleven down. Harbert's great Lab, Raven, goes to a bird, then sees another crippled and swimming for the safety of high marsh grass. . . . By ten o'clock, we have bagged twenty-four mallards, and needed only four more to fill our limits."[10]

There are places like Twin Rivers along all waterways in West Tennessee, but especially the Obion. Whether the hunting is carried out by a club, or from a boat in a patch of willows, one good day is enough to last until the next season. "They'll do better next year" is the motto heard every season. It takes more than one or two, even three or four, bad seasons to keep hunters away from their blinds in the river bottoms. While they talk about the "good old days" of hunting, rarely is it mentioned that the days of hunting among flooded oaks are practically gone.

Water attracts ducks, and thousands of acres along the Obion and Forked Deer have water. But these are artificial swamps that have taken the place of live oaks standing amid free-flowing water, and in these places duck food is scarce or nonexistent. Where ducks have no food, they leave and learn not to return. The trick is to provide food and water at the right place and at the right time. This is possible in those fields where the drainage is good and the crops are not threatened by annual floods and the encroachment of artificial swamps. Successful food crops can be raised in these areas and later flooded for the hunting season. Most duck hunters are aware of this and have begun to adapt.

Their success in acclimating to the loss of the area's native rivers, however, has not been repeated among other hunters—those who go after raccoon, quail, turkey, and deer. These sportsmen have largely disappeared because there is so little forested land in the vicinity of the Obion–Forked Deer complex.

The Fishermen

Fishing for largemouth bass, crappie, bluegills, and catfish remains a popular sport in West Tennessee rivers where fishable water can be found. Typically (and not surprisingly), according to what few surveys have been conducted, the channeled rivers have seen far less use than the relatively natural streams such as the Hatchie. It is not uncommon to find as few as fifty pounds of fish per acre in the Obion or Forked Deer and even fewer pounds per acre in the artificial swamps. In the waters of the Hatchie, however, one can find three hundred to four hundred pounds of fish per acre. Meandering streams and free-flowing floodplains are necessary for good fishing, and channelized streams simply do not provide this preferred habitat (although a few savvy anglers can still find a fishing hole here and there). While such heavily altered waterways lack organisms for food and favorite nooks for fish, the most limiting factor is usually an adequate supply of oxygen. Many of the old fishing holes are now stagnant swamps, long since depleted of the oxygen needed for fish survival.

Studies in the early 1990s clearly showed the poor quality of the artificial swamps. In its attempts to restore the floodplains of the Tigrett wildlife area, the TWRA evaluated places where fishing was known to occur in the Forked Deer River bottoms. TWRA fishery biologists Tim Broadbent and Bobby Wilson found interesting comparisons between an artificial swamp and two different stretches of the old river that had been left stranded by channelization. As one might expect, the artificial swamp had extremely low oxygen, but the segments of the old river meanders—one with some flow and the other with a well-defined flow—had enough available oxygen to support good fish populations and sometimes fair fishing. Samples in the artificial swamp had from two to four parts per million of oxygen, well below the required five to six parts per million needed for healthy fish populations; thus what was once a popular fishing hole no longer supported this activity. This swamp had forty-two pounds of fish per acre with low fish numbers and few species. In comparison, the two segments of the old river had from eight to twelve parts per million of available oxygen. Both sample sites had more than six hundred pounds of fish per acre. Numbers and diversity of fish were also high.[11]

Good statistics throughout most of the rivers are largely unavailable since the TWRA stopped collecting fishing data years ago.[12] Nevertheless, subjective surveys at fisherman-access areas and personal contacts by wildlife area managers suggest that fishing is generally very low in these rivers. The exceptions are in old swamps where aquatic weed control and artificial water-level manipulation occur; sometimes the parking lots at these areas are nearly filled to capacity during prime fishing seasons.

The Farmers and Landowners

Farming is the main activity within the Obion–Forked Deer floodplains, which encompass 759,000 acres in fourteen counties. More than half of this area is in cropland, and 87 percent of this is in soybeans. Fifty-five percent of the state's soybeans and cotton crop, 40 percent of its wheat, 30 percent of its corn, and 35 percent of its bottomland hardwoods are grown here. Sixty-three percent of the annual flooding occurs on 63 percent of the floodplain, and it is here that most of the flood damage occurs. In 1971, for example, the damage amounted to $31 million.[13] Fragile soils that are ten times more subject to erosion than the national average present a serious problem for farmers and landowners in all of West Tennessee's rivers. An estimated 2.5 inches of this fertile topsoil washes downstream every fifteen years in the Obion–Forked Deer complex, five times the amount that can be tolerated without loss of productivity.[14] Thus, special pains must be taken to control soil erosion.

In the 1950s, soil conservationists thought that the soil problems in this region had been conquered. Conservation techniques such as contour farming and cover crops were widely used. Highly erodible land was taken out of crop production, and much of it was reforested. But then, in the 1960s, the high demand for soybeans practically neutralized all of the earlier work, putting land conservation on the back burner. Upward-spiraling costs of farm machinery, government policies to pressure the farmers for higher yields to "feed the hungry world," and requirements for larger, more efficient equipment added to the problem. A significant segment of this farmland belonged to absentee landowners, and this too had its effect on the land. Tenant farmers who worked these lands on short-term contracts tended to ignore sound conservation practices, exacerbating the erosion problems. The Corps of Engineers increased its West Tennessee Tributaries Project with an additional 160 miles of river channels, which encouraged even more land clearing. The project was expected to benefit 140,000 acres of cropland and permit the clearing of 180,000 acres of bottomland hardwoods for agricultural purposes.[15] With this, larger crop acreage and larger fields were demanded. In the flat delta of the Mississippi River, for example, fields went from around five hundred acres to nearly a thousand acres. With bigger farm equipment and larger fields, contour farming and fence rows became a management practice of the past, and soil erosion increased as wildlife habitat decreased.

Landowners depending on the timber industry also suffered. Cropland had replaced 41 percent of the bottomland hardwood forest by 1950. Levee building followed as the soybean decades of the 1960s and 1970s gathered momentum. Twenty-three percent of the original 431,000 acres of bottomland hardwoods had already disappeared in Tennessee; now a larger bite occurred. And while those who converted these areas suffered the consequences, so did waterfowl and other wildlife.

Clearing floodplains for the soybean industry became common practice, and it demanded that these fields be protected from increasingly higher flood regimes. The stage was set for greater problems: The high risk of farming wetlands began to take its toll, and farmers and landowners complained louder than before. With help from the government, bigger canals were constructed, and with these came huge stacks of spoil dirt, which acted as levees. Farmers took advantage of these levees and added more of them. Now the crop fields were barricaded from the river floods, but decreasing the flooding in one location meant that it had to increase at some other place—usually upstream or on the opposite side of the river.

There are at least two problems associated with barricading a river. One is that the flow and the frequency of flooding increases, which leads to soil erosion and sediment deposition. The other is that soybean fields are likely to be found on both sides of the river—and both require flood protection. This means that, eventually, a large portion of the floodplain is compartmentalized with levees that run along each side of the river—parallel levees, in other words. The main flow of the river is constricted and forced through the narrow gaps between opposing levees. This causes the river to rise higher than normal and to back upstream, sometimes for miles. Drainage is limited to the main channel in all but major flood events, in which levees are frequently overtopped.[16]

By the mid-1960s, encouraged by government projects to enlarge the river channels, the use of levees became common practice. In some cases, individual farmers cleared thousands of acres for soybean fields and protected them with levees. State and federal agencies encouraged this work or turned a blind eye to it. It soon became obvious that benefits accruing to one landowner came at the expense of another. Frequently this led to countermeasures—high levees on one side of the river called for the neighboring farms on the other side of the river to build even higher levees. In effect, the government agencies promoted ever-escalating "levee wars" between competing farmers.

Some of those wars ended up in court, as in the *Blackwell v. Butler* case (Obion County Court of Appeals, 1978). A levee and a ditch caused water to back up on the plaintiff. The court declared: "Any substantial or essential interference with flow, if wrongful, whether attended with actual damage or not, is an actionable nuisance." In a similar case (*Adams, Capps, and Herron Farms v. Ladd,* Weakley County Chancery Court, 1983), in which one of the plaintiffs was a state senator, the court ordered abatement of the nuisance without making the plaintiffs prove damages.[17] Thus, it is clear that the courts do not take kindly to those who flood others, though such cases have rarely made the news.

There was simply something wrong about excluding the floodplain from the river at the expense of others. It was obvious that taking up space in the floodplain was bound to cause more flooding problems for someone else. To be sure, farming can be a desperate business, and one's conscience, one's feelings about right and wrong, can sometimes be overridden when it means the difference between a good year and a bad one.

Perhaps many farmers thought it best simply to put up with each other in the hope that no one would notice that there were questionable practices on all sides. But what the farmers who built levees had done was really no worse than what others had done along the rivers. After all, the highway departments and the railroads did not limit their road building and track laying—which required constructing their own versions of levees—to one side of the floodplain; in fact, their structures blocked the river in hundreds of places throughout the Obion–Forked Deer system. For that matter, nearly all waterfowl units along the rivers used levees to protect food crops and hold water on these crops for ducks and duck hunting—a practice that often constricted free-flowing water.

Other Users of the Rivers

While it may sometimes seem that the fate of West Tennessee rivers will lie in the hands of those with large economic interests—large-scale farmers, government agencies, landowners, and well-organized and motivated hunting organizations—we cannot forget the other river users; they also have a stake in the future of these waterways. I am speaking here of boaters, canoeists, bird-watchers, students of nature, vacationing families, and others for whom rivers offer a reprieve from ordinary life. A river can be their refuge. In late winter, they see wood ducks skillfully navigating the lofty sycamores; they hear the birds' loquacious squeaks or mournful squeals and marvel at the eloquence of their quiet splashing at the bend of the river. They find solace in an October walk along abandoned log roads by the banks of the river while they shuffle ankle-deep in colored leaves beneath giant oaks and hickories and fill their pockets with hickory nuts and sweet pecans. They might float the river meanders in canoes or spend a carefree day relaxing on the riverbank. Our urban world is filled with people who desire and need retreats to peaceful places like rivers.

Unfortunately, there are practically no statistics on these users in West Tennessee, mainly because most of the rivers have deteriorated to such an extent that their human use is mostly limited to waterfowl hunting. Canoeists find no pleasure in staring at the raw banks of ditches with few trees on either side. There is little adventure if they portage the high banks only to reach a dead or dying floodplain covered by stagnant water. Bird-watchers encounter a similar problem, finding an extremely limited diversity and number of their favorite avian wildlife.

In places like Reelfoot Lake, however, where good numbers and diversity of wildlife still exist, we find that bird-watching, fishing, photography, camping, and sight-seeing —indeed, all recreational categories except hunting—make up 88 percent of the visitor use of these areas.[18] We still find these types of use very high on streams such as the Hatchie River, which have access accommodations and retain an appreciable amount of their native river characteristics. Here we still find canoeists, hikers, primitive campers, and fishermen—indeed, nearly all outdoor recreationists who require scenic and interesting wetlands.

Chapter 3

The Dynamics of a River and the

Coming of Civilization

A river is eerily close to being a living organism with character, metabolism, and movement. In a continual state of change and aging, it lives and breathes as the sum of its living communities. It often seems to have moods, responding to pampering or abuse in a fashion some might call emotional. Nothing influences the character of the river more than its soils and the climate, except perhaps human beings. In the absence of civilization, the river constantly adapts to the changing environment and ultimately finds a state of harmony or equilibrium; it sweeps along constantly with the continuum of time, renewing itself and thereby sustaining its abundance of natural resources. Even with the presence of civilization, all it really needs is good stewardship: if we use it wisely, it tolerates conflict well, and both people and the river benefit. But to use the river wisely we must first appreciate its dynamics, recognize its benefits, and give it room to flex and breathe. Rivers are born to be wild, and while we may temporarily disable them, we will never tame them.

A river is not simply the peaceful flow we see in July or August, as some might suppose. Take the Mississippi—the mother of West Tennessee rivers. It can be lazy, clean, clear, and placid during late summer and early fall, or it can be unbelievably turbulent and muddy during peak periods of high runoff. Its design is perfectly adapted to the climate and the landscape within its watershed. The Mississippi River watershed covers three-quarters of the contiguous United States, so it has a tremendous influence on people as well on as the land. Conversely, people can have a tremendous influence on the river. The modern misconception that a river is normal only at its low stages is partly to blame for the endless conflicts between rivers and people. Lessons from the Mississippi might give us reason to think differently about the rivers in our own backyard. It might also give insight about why there are conflicts, social restlessness, and unmet needs when equitable land stewardship breaks down. If we hope to continue to benefit from rivers, then we must change the way we perceive and use them.

Catastrophic floods occurred on the Mississippi River in 1882, 1884, 1886, 1887, 1912, 1927, and, most recently, 1937. In comparison, local floods, such as those of 1972, 1973, 1979, 1984, 1985, 1989, and 1997, are minor. The rise and fall of the river stages are key factors in the ecology and dynamics of the river. Passing time and the cyclic effects of droughts and floods on a given landscape form the characteristics of the river—and the wetlands it creates and mothers.

Truly extreme floods, much greater than those mentioned above, may occur every five hundred years or so, perhaps less frequently. Since the period of record keeping is relatively brief, we do not know the extremes of such great floods. Nevertheless, climatologists speak of thousand-year flood events just as they speak of ten-, twenty-, or fifty-year flood events. Thus, we can speculate that sometime before the New Madrid Earthquake of 1811–12, and before the arrival of Europeans, the rains feeding the Mississippi started earlier and lasted longer. The Native Americans living along the river moved out of the floodplain to higher ground along the Chickasaw Bluff. The rains continued off and on through spring, then into the winter, and the next spring. The torrential rain seemed never to stop. It far exceeded the normal fifty-four inches of annual rainfall expected in this region and was likely as severe or worse in the regions of the Missouri and Ohio rivers in the Upper Mississippi drainage.

The Mississippi rose with inordinate rapidity. The currents churned and increased in velocity to speeds of eighteen miles an hour or so. A great torrent of chocolate-colored water rushed over river bars and across the floodplains. The roiling river rose higher and higher until it threatened everything in its path. It caused all manner of havoc. Earth was plowed away in acre-sized chunks; trees snapped like twigs and piled in large log rafts. Logs and other debris stacked against the point bars, and some bars washed away. Whole trees, root wads and all, were torn from the riverbanks and disappeared. Great columns of water boiled from the bowels of the riverbed like underwater geysers. Whirlpools ninety feet deep and a hundred yards across sucked trees to the bottom, not to be seen for miles downstream, or ever again. Sediments were swept from the floodplain, and new outlets were formed. All five rivers along the West Tennessee border were backed upstream for miles with thirty or more vertical feet of backwater in their lower reaches. Finally, the rivers receded. Strong currents gushed back from the floodplains and the tributary rivers.

As the flood waters receded, the Native Americans began returning to their old camps. They waited for the river to fall still further so that they might continue to fish, hunt, and enjoy nature's benefits as usual. They doubtless understood the need for the river to undergo periodic changes and had no complaint.

Over the many years following the flood, giant bottomland hardwoods grew to heights of more than 120 feet. A cathedral of verdant green, a jungle of cane, vines, and other tangle, was created. The topsoil was now 12 feet deep from decades or centuries of build-up. Wildlife flourished in the thick undergrowth and open canopy of the tall hardwoods. Fishes and other aquatic animals filled the creeks, swamps, marshes, and lakes.

All of this was quite natural. Historic floods restore the river, and we are the ultimate benefactors. During such events, heavy loads of nutrients, debris, and sediments are carried to the main channel of the Mississippi and to waiting sea life in the Gulf of Mexico. Runoff provides important nutrients to rivers and wetlands as well as coastal estuaries and the ocean, feeding microorganisms that will provide forage for the larger organisms, building food chains that ultimately produce not only the animal life of the rivers but of the oceans. Channels along the Mississippi are cleared of obstructions, and more efficient channels and floodways are created.

The Mississippi River, during the great flood described above, changed its course here and there. Morphological changes occurred throughout its length. When the river came to a rest, a temporary equilibrium was established, and the river regenerated all that was lost and more. The temporary tranquility would last until the next flood, when the process would be repeated, though not on such a huge scale. In fact, the river would sometimes go to the opposite extreme. During prolonged droughts, it would fall to its lowest stages, and this too would regenerate the river. As nature would have it, the extremes of floods and droughts are never exactly the same. Nature abhors stability.

It often takes more than one great event to produce the magnificent natural features found in rivers. During great floods, some segments of the Mississippi are abandoned and form oxbows—natural lakes. Given time and natural processes of succession, oxbows are filled with water and fish. Eventually, they naturally fill with sediments and become dry ground. Islands form, and willow sprouts invade the new ground the following spring and summer.

Created mostly by aging oxbows, wetlands such as meadows, marshes, and swamps become hatcheries and release fish spawn and other animal life into the parent streams. Some of the young fish leave the oxbows and their residual wetlands with the next rise in the river. Thus, wetlands are a major source of fish to the tributaries. Eventually, some fish travel downstream from the tributary rivers and supplement the Mississippi itself. Over time—decades, centuries, millennia—the cycles of high and low water and the modifications that result from them form the characteristics and interworkings of the river, its ecosystem. Despite events that might appear to us as cataclysmic, the ultimate outcome will be favorable for all creatures, including humans if they choose to accept and promote it to their greatest advantage. In the summer after a great flood, the ecosystem is left revitalized. The floodplain returns to its typical flood regimes, and a fine layer of sediments and nutrients settles over the entire valley.

It was during these great flood events that the Mississippi shifted some twenty miles north of the confluence of the Obion. The new channel formed a large pendulum-shaped tract of land that became Bessy Bend (often called the New Madrid Bend, or Kentucky Bend), the northern portion of which would eventually be in Kentucky. The new bend added some twenty to thirty miles to the river. The switch also left a part of the old river, an oxbow, precisely where Reelfoot Lake now exists between the river and the toe of the Chickasaw Bluffs. This oxbow was, coincidentally, about the same length as Bessy Bend—a lake and swamp that stretched from well into Kentucky all the way to the Obion River in Dyer County, "The Land of Shakes." Later, Reelfoot Lake became the centerpiece of this area. Below the lake were the "Scatters," some fifteen to twenty miles of swampland mixed with hardwood ridges and the sunken channel of the Reelfoot River. The evidence of the Scatters can still be observed in the geological records, in historical accounts like those by Davy Crockett, and in the land surveys of Henry Rutherford and others.

As if to add emphasis to the dynamics of the flooding on the Mississippi, in December 1811 and on into the winter of 1812, a series of earthquakes sank Reelfoot River and the

The Mississippi River flowing around Bessy Bend in Lake County, Tennessee. Reelfoot Lake is the oxbow in the distant background noted by an arrow. (Photo by Jim Johnson.)

cypress swamp, perhaps by as much as twelve feet. The large sink, filled by springs, runoff, and the waters of the Mississippi, covered the cypress swamp and created Reelfoot Lake. Reelfoot River was soon forgotten. Remnants of the river are known only as the Bayou du Chein and Reelfoot Creek in its upper reaches. The tail of the river is now a ditch named Running Reelfoot, supposedly so called because no one thought it appeared large enough to be a river.

Reelfoot Lake at its best lasted for nearly a hundred years, but then civilization arrived in greater and greater numbers. Unlike the activities of Native Americans, the footsteps and ambitions of some white settlers would not be benign. Some came and stayed, often without much thought about the reasons why they came. The bounties of the river and the floodplains—the wetlands, fertile soils, scenic beauty, timber, and wildlife—were taken for granted. These settlers before the industrial age, and those succeeding them, were mostly preoccupied with how to exploit nature's resources for the needs of the day, and in more recent times, we have followed with much the same mindset. Levees were built, wetlands were cleared, and Reelfoot Lake was separated from the river. Tolerance for the normal functions of the river and the wetlands of Reelfoot became virtually nil. The lake gained national acclaim as a natural area of unique biological and historical significance, but only a few locals in the region can remember the magnificence of the Reelfoot wetlands and why they had enjoyed such renown. Why the area changed is left to afterthought. Today we continue to pit our immediate needs against the long-term stewardship that would restore the magnificence of these wetland ecosystems.

The point is that drastic natural events should not only be expected from rivers; these, or some semblance of them, are essential to prevent a net loss and to continue prosperity.

While we might lessen the effects we can no longer tolerate, it is erroneous to think we can prevent them entirely. The wise thing to do is to anticipate them, frame our needs and uses around the inevitable consequence of living in the floodplain, and sustain the benefits we hold in high regard. To do otherwise will cause us to face inevitable destruction, like that which occurred during the flood of 1927, or the creeping and costly burden and decadence we now will see in this short history of West Tennessee rivers.

We should begin by understanding the term "flood zone." A flood zone is often used to describe the various stages of the river—the annual flood zone, the five-year flood zone, ten-year flood zone, twenty-year flood zone, and so on. "Twenty-year flood zone," for example, refers to the level that flood waters are expected to reach, on average, every twenty years. When the river goes higher than about a twenty-year flood zone (sometimes even the annual flood zone), some river user is likely to object. Planners often use the one-hundred-year flood stage as the standard for long-range planning. But some river users (and planners) ignore even the annual flood zone and attempt to raise crops and build roads, houses, levees, or other facilities where the waters are likely to spread. This kind of shortsighted thinking is the root cause of our conflicts with rivers. The annual floodplain has a 100-percent chance of flooding each year. As strange as it might seem, those who encroach upon the floodplains have one major complaint against the river—that it floods and threatens their property nearly every year.

Rivers and Settlements: A Brief History

The problems with the tributary rivers of West Tennessee soon began after the first large settlements. Rivers—often called the "blue arteries of life"—are not always clear and clean or so blue. Even before the European settlers entered the West Tennessee country, the tributaries were probably what the early adventurers would have called "muddy" during seasons of heavy rainfall. This appearance was the result of the fine loess soils in the region, which are very unstable. But stream "clarity" is a relative term. If inhabitants of two hundred years ago saw the rivers today, they would have to develop a completely different understanding of "muddy water." The term "Muddy Mississippi" was probably not a term used before Mark Twain's era. The Agonquin Indians called it the Misi Sipi, or Big River. Hernando DeSoto called it the Great River and the Father of Waters. Jacques Marquette spelled it "Mitchisipi."[1] But only the modern descriptions of this river characterize it as "muddy." Even today, during October droughts, the Mississippi runs clear, just like its West Tennessee tributaries, but it is rarely clear during the rest of the year. We would still see these streams running relatively clean and clear if it were not for the disturbance caused by the human failure to observe good conservation practices.

The relationship between rivers and people was different less than two hundred years ago. People needed everything the river had to offer—sustenance, and prosperity, relaxation, and for some, spiritual rest. Rivers were also highways for people and wildlife. These attributes of rivers are similar for people today. But Native Americans and the earliest white settlers living in the floodplain learned early to plan—to respect the river, to be mobile, and to adapt to the inevitable challenges and changes. Those who stayed in the floodplain during floods lived on piers or boats, a common-sense practice that many have since forgotten.

Few who lived near the river were likely to describe it as boring. If something curious was not floating downstream, fish and wildlife and people were going upstream. For those who lived along the river, much of their thinking and way of speaking revolved around it: something was either "up the river" or "down at the river" or "at the river." For such inhabitants, the river was nothing less than a part of their very being. It did not preclude their being ambitious; survival required ambition then, as it does now. But being ambitious is never an excuse to leave the land worse off than one finds it.

In the late 1700s and into the early 1800s, settlers such as Davy Crockett found the rivers of West Tennessee in a pristine state. Then, the land was sparsely populated; it is rare to find any descriptions of life around what became Reelfoot Lake before it was created. The watersheds were as fragile then as today, but there is enough evidence to show that the rivers were relatively clean and the disturbances to them minimal. People like Crockett would find it impossible to imagine the present condition of these streams.

Navigation reports, surveys, and maps from those early days indicate that rivers such as the Obion, the Forked Deer, and the Hatchie were discrete and often deep enough to accommodate fairly large boats for transporting goods. Stands of mature forests were still present, especially in the lowlands of these rivers and the Mississippi River delta. Typical was the forest found along the Hatchie and around Reelfoot Lake today. Diversity and abundance of wildlife—waterfowl, big and small game, wetland and terrestrial species, resident and migrant songbirds—were evident. Even today, the greatest abundance and diversity of wildlife are found in the remnant forests along the rivers.

The Early Settlers

In the 1700s, settlements were relatively small and scattered in the region, with negligible impact on the native rivers. Memphis and Jackson were not established until 1819. These were followed by Dyersburg in 1850 and by Tiptonville and Samburg in 1857. Surveying the area as early as 1753, however, was Henry Rutherford. In what would become Dyer County, Rutherford established a reference monument, or key reference, for all surveys in the region. Thus, the community that developed at this site became known as "Key Corner." The site was isolated from Dyer County in 1831 by the ever-changing river, and it became part of Lauderdale County in 1835.[2] By this time, small family farms were scattered here and there, east of the Chickasaw Bluffs, where farming was profitable in the fertile hills of the watersheds. The effect of civilization on the rivers was just starting to be felt. Hunting big game and trapping for furbearers were important enterprises, especially between the bluffs and the Mississippi River. Eventually, roads were built through most of the country, and riverboats established trade on the rivers.

An active river trade along the Obion and Forked Deer rivers developed during the early 1800s. The North Forked Deer was navigable for small steamboats as far upstream as Dyersburg for about nine months of the year. Boats ascended the South Forked Deer River as far as Jackson. During certain seasons the Obion River was navigable for the lumber industry up to the town of Rutherford.[3]

Earl Willoughby has written about some of these early trading practices, and his research gives insight into the status of the local rivers during those days. According to Willoughby, Barney Mitchell became the first merchant in 1818 to establish regular trade routes to Jackson, some ninety miles up the Forked Deer River. By 1822 merchant boats (which included keelboats that could be poled upstream or downstream and flatboats that went one way, downstream) were coming to the tavern of Colonel Robert H. Dyer on the North Forked Deer. Merchant boats also went up the South Forked Deer River to the place of a Dr. Butler near Jackson, some sixty road-miles from the Mississippi. "By 1823, settlers were spreading throughout West Tennessee in great numbers," Willoughby noted. "Most of these people came down the Mississippi on keelboats and flatboats. The rivers were alive with activity."[4]

A steamboat named the *Red Rover,* according to information in the Haywood County archives, is said to have been the first of its kind to travel the Hatchie River.[5] In 1825 the steamboat went upstream as far as Bolivar. A steamboat could not make the first five miles of the trip today. It is also doubtful whether flatboats or keelboats could ply the shallow Hatchie or any other of the five rivers. Because of sand accumulation, one would even find it difficult to run a small, narrow, shallow-water johnboat on these rivers during normal midsummer river stages.

In his 1834 autobiography, *A Narrative of the Life of David Crockett of the State of Tennessee,* the renowned frontiersman offered a lively and probably accurate account of what settlers were likely to encounter along West Tennessee waterways of nearly two hundred years ago. In 1822 Crockett returned from the legislature to his home on the Rutherford Fork. "I took my eldest son John with me and a young man by the name of Abram Henry, and cut out for the Obion," he noted. According to Crockett, the country was a wilderness with a tremendous number of bears, deer, and wild turkey, as well as a fair number of panthers and elk.[6]

On this trip Crockett and his party went directly to Obion Lake, where he intended to build a log cabin. This site is today probably at Round Lake located between Lane and Sharpe's Ferry. Crockett's party soon discovered a large keelboat moored at the other side of the lake: "[M]yself and my young man went to the boat with Mr. Owens [his nearest neighbor living seven miles away on the opposite side of the river] and the others. The boat was loaded with whiskey, flour, sugar, coffee, salt, castings, and other articles suitable for the country; and they were to receive five hundred dollars to land the load at M'Lemore's Bluff [located between present-day Trenton and Dresden]." Crockett had already decided that he needed to build his cabin in a prairie-like opening near the lake. "And so I got the boatsmen to go out with me to where I was going to settle, and we slap'd up a cabin in little or no time," he wrote.[7]

Crockett later decided to try his hand as a river merchant: "In the fall of 1825, I concluded I would build two large boats, and load them with pipe staves for the market. . . .

I worked with my hands until the bears got fat." The country was abundant with game big and small during those years. Crockett and his party killed fifteen bears in two weeks, and then went on to kill several more in the Obion and Reelfoot Lake vicinity. Before finishing the hunt, the party killed a panther and six elk.[8]

In late winter of 1826, Crockett returned to his boat-building and stave project. "I had two boats and about thirty thousand staves," he recalled, "and so I loaded them and set out for New Orleans. I got out of the Obion River . . . very well; but when I got into the Mississippi, I found all my hands were bad scared, and in fact I believe I was scared a little worst of any; for I had never been down the river, and I soon discovered that my pilot was as ignorant of the business as myself."[9]

Crockett gained a new respect for the river and its behavior. His expedition never made it to Memphis, much less New Orleans. A few miles above Memphis, in a treacherous bend of the river called Devil's Elbow, his enterprise was wrecked. The turbulent undercurrents of the Mississippi engulfed the two boats and their contents. Crockett escaped with a good portion of the skin stripped from his body. The only way out of the boat's submerging cabin was through "a small hole in the side, which we had used to put our arms through to dip up water." Crockett continued: "I put my arms through and hollered as loud as I could roar . . . and the hands who were next to the raft, seeing my arms out, and hearing me holler, seized them and began to pull. . . . By a violent effort they jerked me through. . . . I was literally skinned like a rabbit."[10] Crockett would probably be fortunate today to make the first leg of his journey. Because of the heavy sand deposits resulting from poor soil conservation in the upper watersheds, the Obion now becomes so shallow that Crockett's loaded boat would be grounded.

Crockett encountered little human habitation beyond the vicinity of his cabin on the banks of Obion Lake. He did, however, mention a few inhabitants, such as a hermit on Reelfoot Lake, and he frequently glimpsed Native Americans on his hunting expeditions.[11] The coveted lands in the region, including the Reelfoot Lake area, had once been claimed by the Chickasaws, but following a treaty negotiated with the tribe in 1818, the lands of West Tennessee were opened for settlement.[12] Both the native peoples and early frontiersmen like Crockett were no doubt aware that the wetlands in and around places like the Obion and Reelfoot Lake were more than simply "unhealthy swamps," as many would call them before they were drained and exploited. They understood the capability of the land, and such early accounts as Crockett's offer valuable insights that modern-day users and land managers would do well to take to heart. During his sojourns in the West Tennessee wilderness, Crockett routinely traveled fifteen to twenty miles from his cabin to hunt in and around Reelfoot Lake, staying for several days or weeks at a time. In one instance, Crockett "found bear very plenty, and, indeed, all sorts of game and wild varmints, except buffalo."[13] He knew that one of his favorite game—black bear—preferred habitat like large trees, canebrakes, and the regenerated growth created by violent acts of nature like tornadoes. Once the trees were thrown to the ground, the forest floor was exposed to sunlight, which started the natural succession of the forest all over; berries, persimmon, grapes, small animals, and other life would flourish, providing ample food for the black bears. "I was sure there must be a heap of bears in the fallen timber," Crockett speculated.[14]

Crockett and his party killed four bears on one hunt. Along the way (probably in Glady Hollow on Reelfoot Creek), he ended up killing a wounded bear—with his knife, no less. Interestingly enough, Crockett mentioned that when the bear tried to escape, he went into one of the many cavernous "cracks" said to have been caused by the earthquake. This proved to be the bear's undoing, for here Crockett and the animal had their final encounter.[15]

Crockett apparently gazed across the waters of Reelfoot Lake. One of his biographers, Constance Rourke, wrote that Crockett "skirted the great cracks still left by the earth quake, crossed low streams, and found Reelfoot Lake with its waters aglow with the yellow lights of the great lilies [American lotus]." It is interesting to know that the American lotus, a pond plant, was present even then, suggesting that Reelfoot was an independent pool soon after the earthquake that had created it. Rourke writes that once, at the head of the lake (probably Upper Blue Basin), Crockett and his son passed a little thatched hut where they caught a glimpse of the old hermit who lived there. "We won't bother him since he wants to be by himself," Crockett said. "I know that feeling well."[16]

Crockett claimed that he and his party killed 105 bears during one fall, winter, and spring.[17] This remarkable number suggests the potential wildness this land still has to offer.

Tennessee's Last Frontier

As wild as the country was, civilizing notions had preceded Davy Crockett to the watersheds of West Tennessee, though not by much. Tennessee was admitted as a state in 1796, and the western district was in Chickasaw possession even before this time. Not until 1818 did Tennessee establish its present boundaries. The country in the western part of the state was more or less unsettled—Tennessee's last frontier. Notable settlements did not happen here until Issac Shelby, ex-governor of Kentucky, and General Andrew Jackson negotiated their treaty with the Chickasaws at Old Town, Chickasaw Nation, on October 19, 1818. With this purchase, claim and title to the land between the Tennessee River and the Mississippi River were ceded to the state for the bargain sum of three hundred thousand dollars, and the greater part of West Tennessee was settled rapidly—all except for the Reelfoot country in the northwest part of the state. It was well into the century before land grants to certain military officials for former military service opened the way for settlers to homestead in the area around Reelfoot Lake.[18] Many had settled earlier along the Tennessee and the upper watersheds of the Mississippi tributaries for fish and game, furbearers, lumber, fertile land, transportation, and trade. Strategically and conveniently located near rivers, the settlements became the towns and cities we have today. Within the period of one generation, enough of the forest was cleared for agriculture to create serious soil erosion problems, and West Tennessee rivers began to feel the effects.

Conserving natural resources had little meaning for some early homesteaders. When the land was depleted, they often gathered their belongings and moved to new and more fertile places, a few miles or a few days away. Recovery of the depleted land and the natural resources was left to others, and where the land was not renewed, it often eroded to gullies.

The activities of civilization moved farther west at about the time Davy Crockett made his enterprising trip down the Obion with his load of staves. The timber industry had already picked up momentum, thus impacting the heavily harvested watersheds. In only a few short years, the giant hardwoods of the river bottoms along the Obion, Forked Deer, and other West Tennessee tributaries yielded to the ax and the crosscut saw. An industrial mindset that promised more civil and comfortable ways of life established itself among the new arrivals.

Storekeepers, doctors, shoemakers, blacksmiths, and others followed the timber cutters. Together, small farming operations and businesses gave rise to growing communities, first in the upper watersheds and then in the river bottoms. The first settlement around the Dyer County area, according to the county archives, was in 1820 on the banks of the Forked Deer River. After this, the towns of northwest Tennessee were rapidly established, and these too were near the rivers. A bustling economy was underway. It took another thirty years before the Mississippi River bottoms of Dyer and Lake counties were attractive enough to encourage urbanites.

The settlers began to hear better accounts of the extraordinary natural resources around Reelfoot Lake. In his 1915 book, *Reelfoot Lake Fishing and Duck Shooting*, R. E. Lee Eagle explained why the early years were so attractive: "The health [of Reelfoot Lake] is as good as you will find at any point on the Mississippi River. The farming land adjacent and surrounding the lake yields a bale of cotton per acre, and sells on the market for one hundred dollars per acre cash, which has a rental value of eight dollars to ten dollars per acre. . . . duck shooting is good and fishing is at its best."[19]

He also wrote: "Over four hundred natives have grown up with their families, and are now using the most modern tackle fishing for profit, with over a hundred thousand dollars invested; taking from the waters of Reelfoot Lake an average of ten thousand pounds of fish daily; shipping to New York, Chicago, St Louis, and other larger cities, besides inland towns and supplying the local trade."[20]

Towns and cities north and south along the Mississippi River thrived, since the river offered a way to transport raw materials and goods. This was an age of engineering and of determined and fiercely competitive men who sought to master the land. Throughout the middle third of the 1800s, one of the most intense competitions involved the drive to subdue and control the Mississippi River. A key incentive for this effort was the need for larger river barges with more load capacity. Such barges required deeper river channels, and the most logical alternative was to deepen the existing channel. And so channelization, though not a common term in those days, began. While increasing transportation efficiency and short-term profits, this effort and its aftermath greatly affected the five tributary rivers in West Tennessee.

This early history is not only interesting and often colorful, but it is also pertinent to the management of West Tennessee rivers and wetlands today. Only by knowing this history can wetland managers gain the insight needed to truly and confidently develop good stewardship plans and practices. Historic accounts enhance one's understanding of what the biology of these river systems were—and how and why they changed. They help explain the evolution of encroachments and where too many compromises were made. One hopes that, armed with this understanding, we will be better able to minimize the pitfalls, educate the public, and build solid foundations for the development and implementation of future plans for the preservation of river ecosystems.

Chapter 4

The Genesis of the Rivers
and the Beginning of Channelization

The watershed is the genesis of the river; it can also be the end for the river. In West Tennessee, the top of the watersheds is the Mississippi-Tennessee river divide. Here the western drainage separates from the rest of the state. The divide runs from north to south, beginning a short distance above the Kentucky-Tennessee border at the head of the Obion River. It trails south more than a hundred miles on an irregular path, following the highest ridge through Natchez Trace State Park, then to the hills east of Jackson, and on to Selmer. Finally, it arcs southwest around the Hatchie and Wolf into the sandy ridges of Mississippi. The entire watershed is formed in a dissected plateau characterized by wide sinuous valleys, gentle rolling hills, hardwood forests, and swamps. The elevation of the plateau at the watershed divide is approximately seven hundred feet above mean sea level. The slope descends some four hundred feet to the western face of the Chickasaw Bluffs. Here, the elevation of the land descends another three hundred to four hundred feet, cutting through bluffs. After eighty miles or so, the channels of the tributary rivers continue variously from ten to twenty miles through a wide, fertile delta, where they are received by the Mississippi River.

The problems for West Tennessee rivers started in the watersheds with soil erosion. It was only a short time after settlers cleared the forests for farmland that erosion became a problem, not for the highland farmers but for the farmers along the rivers. Driven by the exhilaration that comes from the smell of newly turned earth, the early settlers overdid it. Thinking that land was in endless supply, they left the raw earth exposed to the elements too long and too often. Tiny rills began to collect the runoff and carry it to the larger drains and branches. Finally, the mud-laden water would reach the main channel of the rivers, and here the serious problems really began.

Moving water reaches a certain limit at which it can no longer carry suspended sediments. At this point, the sediments settle to the bottom of the channels—first the heavy sediments, then the lighter sediments. As a result, the streams widen, their gradients become increasingly shallow, and the moving water loses energy. The stream channel itself reaches a capacity where it can carry no more water, and it spills over the banks and spreads across the wide floodplain. More energy is lost as more sediments accumulate.

As the channels and their lateral drains become filled with sediment, flooding worsens and affects cropland, pastures, and other human developments—indeed, everything in the floodplain. With this sort of flooding, commerce and travel in West Tennessee were affected. Keel boats and flat boats used to transport people and products could no longer navigate the rivers. A crisis was underway by the late 1800s. By then, railroads had connected the timber resources of the South to markets in the North, and land clearing was rampant. Large-scale dredging campaigns were demanded by the 1920s, and the rivers began to change.

The Thin Blue Lines in a Healthy Ecosystem

The present-day watersheds of West Tennessee are mostly open land—fields, highways, yards, and artificial hard surfaces conducive to rapid runoff. In a sustainable, natural river ecosystem, however, the watershed is largely covered with vegetation, and understanding how such ecosystems function in the best of circumstances will help us to better comprehend what has gone wrong.

A river has about a half-dozen major components: a watershed, the rills and rivulets (the first small streams), branches, creeks, and the valley, which contains the floodplain and the main channel. Flowing from the upper watershed, the branches of a river eventually converge to form a discrete main channel downstream. The patterns of a river—those thin blue lines seen from the air or depicted on a topographic map—are similar to those of a large sycamore leaf: the upper extremes of the watershed are like the outer margins of the leaf, and the streams and branches are like the delicate outer leaf veins, which descend on the central vein from either side.

In a healthy watershed, the raindrops land softly on the leaves of trees and plants and on the spongy layer of decaying vegetation that covers the ground—the "litter." The water at first seeps slowly into the thirsty soil, hardly displacing or disturbing any of the soil structure. Some of the rainfall is used by soil organisms and some of it by the vegetation. With enough rain, the biota and the soil in the spongy top layer become saturated, and the excess water may evaporate and enter atmospheric water cycles, or it might percolate into underground aquifers.

At some point, the watershed becomes saturated enough to be fully recharged. The remaining water leaves the upper watershed as runoff. In ideal conditions, such as nature would have it, runoff flows twisting and turning over and through the soft litter, losing its energy nearly as fast as it is gained, until it reaches the first discrete rivulets. These small streams run clean and clear. This is because they contain very few soil sediments, owing to the watershed being protected and held in place by living vegetation and the litter. Once the runoff collects in the rivulets, the kinetic energy of the stream increases. But then the small stream begins to meander in sharp twists and turns until its velocity reaches a state of equilibrium. With this, the beginning of the river is revealed, and topographers record it as a thin blue line on maps.

As rain continues, the volume of runoff increases, and the various blue lines converge and become progressively larger downstream until they are no longer defined as minor branches or creeks but as the main channel of a river. The river begins to compensate for its increased volume and energy. It continues to stretch, and the blue lines become lazy S's and sharp bends. The river channel also becomes deeper and wider to dissipate the energy and slow velocity, adjusting to reach its optimum capacity and equilibrium.

Finally, the main channel of the stream reaches its threshold, and the overflow spills first into the secondary channels and then onto the floor of the floodplain. This process refurbishes the river valley with nutrients and water. Trees and other vegetation grow densely in the floodplain, constantly resisting the energy of the floodwater flowing downstream. The vegetation mitigates the destructive potential of the floods, holding the life-giving water long enough for it to be absorbed into the rich soils. The excess runoff continues its journey, and the thin blue line becomes much more conspicuous as it converges with an even larger river—for example, the Mississippi. Then it flows to the coastal wetlands and the Gulf of Mexico.

At this point, the entire river is in equilibrium with its environment, sustainable and predictable enough that humans can judge what constitutes safe and equitable use and what does not. In natural streams, equilibrium is met when the energy of the river does not exceed the capacity of the floodplain to tolerate or to adapt to flooding. In the event of natural catastrophic changes, the river will, over time and free from human intervention, adjust its bed and banks to strike a balance with these exceptional events. This is an ongoing process: when not interrupted by civilization, the floodplain continues to evolve and adapt so that flooding does not deteriorate the river ecosystem beyond its capacity to recover; thus the dynamics and vitality of the river are sustained.

The Thin Blue Lines after 1920

A good atlas-gazetteer is one of the most important tools a wildlife manager has if a helicopter or a small plane is not readily available. He will loan out almost all of his possessions except for this resource. Anyone with an interest in the subject can learn more about a river from a "topo" map than from any other tool, and for the waterfowl/wetland manager, it is irreplaceable. It is used daily for navigation, location, planning, and information on the history of the land. Following its blue lines gives us an aerial perspective; supplement it with "ground truth," and the facts about the river come into clear focus.

Once the rivers of West Tennessee began to fill with sediment, local interests during the 1920s and 1930s sought to straighten the natural channels on the Obion–Forked Deer rivers, and the maps of West Tennessee began to show artificial rivers. The theory behind this strategy, following the U.S. Army Corps of Engineers' example on the Mississippi, was that a shorter main channel increased slope, which increased the energy and velocity of the stream—enough to scour the channel. Heavy sediments, then, could be transported downstream out of the river system and out of mind.

With this flawed theory in place, West Tennessee entered the age of channelization. When the lazy S's of a natural stream lose their sinuosity, it is almost a sure sign that man has had a hand in the design. A paddler plying a segment of a natural river constantly strains to anticipate the view around the next bend, and years ago such turns occurred at no more than a hundred yards or so in West Tennessee rivers. But now the river channels and the blue lines on the map show long tangent stretches—zigzags that cross the topography of the river valley like the path of a ricocheting bullet. It is not difficult to see this sad situation from the ground.

James McClinton, a catfisherman on the Forked Deer, was one who noticed. He did not need to float the river (and very few did) to see the changes; he could easily survey them from his perch from the riverbank. "I've never seen a river run straight like these . . . that long, that far," he told me. "They are usually crooked like a snake."[1] McClinton had only the Hatchie in this region for a comparison, where, incidentally, catfishing was remarkably good.

Throughout the length of these rivers, the topographic map shows expanding swamps, encroachments, and wide-open spaces along the corridor. Segments of the old river lie

This skeleton of a cypress tree, more than a hundred feet tall, was found along an old river meander on the Forked Deer that no longer functions, having been trapped in an artificial swamp and bypassed by the river channel. (Photo by Jim Johnson.)

The Genesis of the Rivers

detached alongside the new channel. The old river segments are frequently easy to detect, although they lie listless with no useful function for drainage. The hardwoods that formerly grew along the banks now lie prostrate and decaying beneath the artificial swamps, unable to withstand the effects of permanent flooding. Dead snags—usually of giant cypress—line the banks of the old river scars, as if their purpose is to mark them. For many years after they are cut off from the main channel, the old river segments may be navigable for several miles, sometimes even serving as good fisheries. But with time they stop functioning as river channels or good fishing spots. Most of them receive fresh water only during floods and cease to flow when the river falls.

Without a good map, it is often difficult to discover a riverbank perch like the one James McClinton found; most people are able to gain access to the rivers only from the highway. Even then, access is limited. Rarely is a visitor to one of these rivers able to walk to its banks (which generally consist of excavated piles of earth) or to launch a boat, since swamps are usually found on either side of the channel. At best, the visitor is likely to get only a brief glimpse of the river.

A perspective from the highway is usually misleading. Although James McClinton could view the long straight stretches of the artificial river, he could not see beyond the trees lining the riverbank. Without seeing certain indicators—snags or dead treetops, changes in the forest from oaks and hickory to the more flood-tolerant black willows, river birch, and water maple—one might even judge the river to be in good health. The most damaging effects from the footsteps of civilization are often concealed by those rows of trees. Numerous minor ditches and miles of levees lie unrevealed without a good map. And many ditches and levees—the constant companions of channelization—may not even be on the map, since they are frequently built or redirected by landowners.

What is conspicuous on the map are the vast areas of open space with few trees. These represent land cleared for cotton, corn, and soybeans. These open areas have expanded significantly in the floodplains of the Mississippi River since the mid-1800s. Prior to this time, nearly the entire alluvial valley was a bottomland hardwood forest—one great wetland. For miles on either side of the river was forested land with open glades of grassland. Now there is little forested land within the floodplains, except for those aforementioned trees lining the rivers. On the map, narrow strips of pale green tag alongside the thin blue lines. But surrounding them are vast white patches indicating the open fields that extend far into the surrounding valleys and hills. The creation of these treeless spaces was a gradual but steady process. When the fertile highlands were exhausted and farming increased, agriculture moved to the floodplains, which became even more attractive to the soybean industry of the 1960s.

Crops and wetlands can complement each other, but they can rarely mix. They certainly have not mixed well in and around West Tennessee rivers. Now, in areas not used for farmland, there are artificial swamps that consume space that might otherwise serve as habitat for waterfowl and other wildlife. It is easy to conclude from the map that nearly every aspect of these rivers along the Mississippi has suffered from stress and decline.

The severity of the damage caused by stream modifications often does not show up as a crisis for several years. Our wildlife management areas along the Obion–Forked Deer rivers reached a critical stage well before the 1970s. Eventually it became worse: new lands expected to become WMAs were added, and these suffered the same problems. Today the TWRA manages nearly 65,000 acres of state-owned wetlands in northwest Tennessee for public use. With a change in mindset and some luck, these can perhaps be salvaged. But the real potential lies in restoring nearly a thousand miles of floodplains in the narrow valleys of West Tennessee rivers. Most of this land belongs to private interests—all of whom have their own reasons and methods for changing the rivers. Conservatively, an estimated 640,000 acres of river floodplains had been wrecked beyond profitable use. The good news is that these lands can be restored. If we can clean up after hurricanes, we can restore rivers—and rivers are our inland coastlines, our scenic waterways, and, for many, they are a source of livelihood and contribute immeasurably to the quality of life. Sensible, ethical management, combined with the public will, can do remarkable things. A thousand miles of river—considerably more land than all the state and federal wildlife areas within the region—can be difficult to assess, to say the least, but not out of reach.

Channelization: The "Solution" and the Problem

Maps of West Tennessee—not to mention descriptive accounts like this book—refer to all major streams in the region as "rivers." The term can be misleading, however, for it is doubtful that a channelized stream can truly be called a river. The *Random House Dictionary of the English Language* defines a river as "a natural stream of water of fairly large size flowing in a definite course or channel or series of diverging and converging channels." The key word here is "natural." Not a single stream in West Tennessee fits this definition, for all had suffered involuntary metamorphosis by the 1970s.

The credit—or blame—for channelization should probably go to General Andrew Atkinson Humphreys, who in 1866, as the first head of the U.S. Army Corps of Engineers, conducted an intense project to resolve the navigation problems of the Mississippi River by deepening the channel.[2] Channelization was typical of the industrial mindset of the post–Civil War era. Mastering nature and forcing it to comply with humanity's will was considered one of the highest goals of the government and industry in America of the time. It was the age of the industrial pioneers, and these elites were determined to control the natural resources of the earth—including the wind, raw ore, timber, oil, and, not least, the mighty Mississippi.[3]

The goal of Humphreys and others was to deepen the river enough to accommodate barge traffic from New Orleans to St. Louis. Creating shorter channels, levees, and rock jetties or dikes was considered the key to solving the problem. The logic was that a straight channel cut across the meandering S-curves, or bends of a river, could produce a deeper channel. Falling to the same elevation over a shorter distance increases slope, and greater slope with the same volume of water increases the velocity of the current. Speeding up the current can, in turn, generate enough energy to gouge and scour sediments from the riverbed and thus deepen the channel.

One of the greatest feats in "taming" the Mississippi was to force the needed volumes of water through the shortcut channels. To increase the volume of water at high stages, the river was "squeezed" by large levees offset from the riverbank on either side. The levees and rock jetties, designed by the corps's James Buchanan Eads, pushed the flow of the river at low stages even more to better define the river channels. As the river became shorter, the slope of the riverbed increased, velocity increased, and the channel scoured ever deeper. The levees and rock jetties could be built along any segment of the river. The corps used these hydraulic techniques innumerable times with seeming success, at least initially. Today, hardly a mile of the Mississippi River can be found where levees have not been built to disconnect the floodplain from the main channel.

The Mississippi River channel was scoured by as much as twelve feet along the western border of Dyer County. It began to affect the tributaries of West Tennessee almost immediately. A differential of twelve feet—the descending slope between the tributary outlets and the deeper channel of the Mississippi—causes "head cutting," or an eroding away of the stream bottom. This process migrates upstream: with every mile of head cutting, a mile of river deteriorates. Riverbanks cave in, and the channels become wider and shallower. On the Forked Deer River, head cutting has progressed some thirty to forty miles upstream, a little beyond the city of Dyersburg. Hydrologist Tim Diehl of the U.S. Geological Survey voiced concern in the 1990s that head cutting could cause the sand formations beneath the riverbeds to be exposed and that this, in turn, could cause a cataclysmic collapse of the river channel.[4]

Humphreys's success and that of his contemporaries no doubt encouraged some drainage districts from the 1920s forward to dig artificial channels to control the rivers in West Tennessee. Dig they did, but they could not duplicate the successes enjoyed by the corps's early work on the Mississippi. In fact, the results were just the reverse. The new ditches soon filled with sand, the river hydrology became more chaotic, and flooding was worse than before. Eventually the exorbitant and ongoing costs became more than landowners could bear. They eventually petitioned politicians for federal and state assistance.

Later proponents of channelization—the Farm Bureau, the Soil Conservation Service, the large agricultural concerns, and others supporting the farming industry—complained that the early attempts at channelization never had a real chance to succeed. They argued that continuous channelization was the answer. If the state provided the appropriate funding—enough equipment, enough manpower, and enough endurance—channelization could resolve the flooding problems. This subdue-and-control mindset would assert itself in the most dramatic form yet when the state requested the assistance of the U.S. Army Corps of Engineers.

Nearly a century after the first channelization work, in 1961, the Corps of Engineers' West Tennessee Tributaries Project (WTTP) dug its first straight-line ditches. The corps believed that these rivers could be sculpted and controlled with the same effectiveness that had been seen on the Mississippi River. The corps had reason to feel confident: after all, it had controlled the world's third largest river. The Tennessee General Assembly formed and authorized the Obion River Basin Authority (later the West Tennessee River Basin Authority) to make sure that the corps's work was carried out.

The proponents of the WTTP were probably right that enough money and energy could keep the ditches functioning for their stated purpose, but how much money would be spent and to whose benefit? Many farmers had already seen their soybeans laden with mud and sand, and some saw them washed away. In the end, channelization was not good for anyone. It certainly was not good for farmers—those who depend upon good drainage and resources that only the rivers could provide.

The subdue-and-control approach to rivers has been around ever since. The effects of this so-called progress are still measured by the tonnage of products transported, by flood control, and by the enormous agriculture production in the floodplains along the Mississippi. While it has indeed been a great success story for the relative few, a full assessment has yet to be made on the Mississippi River. On the surface, government economic analysis has shown remarkable returns, but the impact from this "progress" goes much deeper than the river channel or the criteria of river trade and agriculture. While the dredging campaign promoted flood control, the result was the reverse: more flooding occurred than ever before, farmers lost cropland, and worse yet, the state lost its native river ecosystems. We have not begun to calculate that economic loss, not to mention the intangible social costs that include quality of life. What effect have modifications of rivers had on migrant birds, water supplies, weather, and outdoor recreation? What is the cost to replace a river ecosystem? To guide us through the next century, more foresight and practical, comprehensive, interstate land-use plans are clearly in order.

Chapter 5

The Wetland Managers and Their Work

Realizing by the 1980s where the use of the river floodplains was leading us, those of us directly responsible for public wetlands had to change our attitudes about the way they were managed. In a sense, the change was forced upon us, but we were ready and willing to accept it, given that native wetlands were disappearing at an alarming rate. We had originally relied solely on our own ingenuity to deal with the problems, but our method—creating artificial wetlands—did not admit to the trap of depending on what we could do for nature instead of what nature could do for us. Artificial wetlands began to deteriorate as soon as they were built—in stark contrast to native wetlands, which can sustain themselves as orderly ecosystems. It was physically and economically impractical for us to think that we could address the needs of wintering waterfowl, not to mention those of other wildlife, by building artificial wetlands alone. Certainly, we could not recover the benefits of native wetlands or meet the rising demand for more public outdoor use.

Assessing the health of wildlife gave us the first red flag: if wetlands were not suitable for wildlife, they were in all probability not suitable for people. The health of wildlife often serves as an index to the health of the land, and this is particularly true for the ultrasensitive wetlands. At least we knew that when the health, mix, and number of the endemic wildlife were good, the wetlands were likely to be in good shape as well. We found, however, that abundant and healthy wildlife were the exception rather than the rule in West Tennessee, despite our best intentions to improve the lands for waterfowl and duck hunters. While we had failed to provide a balanced management program, we came to see that this goal might be attained if we changed our thinking and could generate the needed support for putting that new thinking into practice.

The Challenge to Managers

Public land for waterfowl hunting and other forms of outdoor recreation has been severely limited since the 1950s, not only in the northern production grounds where waterfowl produce their young but also throughout the winter grounds in the Mississippi Flyway. The situation grew worse after the soybean-production boom of the 1960s, and it continues through the present. "Unfortunately," Scott Yaich pointed out in a 2004 magazine article, "the growth and complexity of society are exerting ever-increasing pressure on North America's finite water resources. The United States has already lost 115 million acres of wetlands—more than half the original total—and continues to lose more than 100,000 additional acres each year. If wetlands continue at that pace, waterfowl are destined to face the equivalent of permanent drought conditions on many of their most important breeding, migration, and wintering areas."[1] Fifty-nine percent of the 2 million acres of wetlands Tennessee once had were lost by the 1980s—leaving a remainder of 814,000 acres—and an estimated 89 percent of these were in West Tennessee.[2] The various ways of modifying the landscape—wetland drainage, urban expansion, highway construction, land clearing, and the like—have been the major causes. Sadly, wetlands are as often despised as admired. Among land developers, the prevailing opinion has often been that the best wetland is a drained wetland. West Tennessee may well have suffered the worst of this trend, but the deficit still climbs.

Managers throughout the Mississippi Flyway have long agreed that unless more public lands are managed for waterfowl, hunters and other wildlife enthusiasts will eventually have no place to enjoy these natural resources. During the 1960s, the best the state could do to correct the problem in West Tennessee was to have the Tennessee Game and Fish Commission (the TWRA's predecessor) acquire land for three small wildlife areas along the Obion and Forked Deer rivers: Moss Island, Gooch, and Tigrett. But then, until the 1990s, the priorities for wetlands were pushed aside since the state had no means to acquire more lands. However sincere the efforts that *were* made may or may not have been, they amounted to little more than a Band-Aid.

Still, whatever their insufficiency, the lands purchased in the '60s and the efforts to maintain them did help to create and legitimize a relatively new kind of state agent: the waterfowl manager, whose ultimate constituent was the public and whose responsibilities would eventually come to include all endemic wildlife—indeed, the whole of the wetlands themselves. The faces of these managers might change over the years, but their stamina, commitment, and dedication have been constant.

Waterfowl managers became known as the "duck heads" in the state wildlife agency. When given the freedom to do their jobs, they have been out in front, anticipating the obstacles, trying to avoid political pitfalls, and forever thinking about better ways of dealing with new initiatives for managing wetlands. Their assignments are rarely easy, requiring constant and intensive management to meet their objectives before and during each waterfowl hunting season. Also, they feel continual pressure to appease politicians, hunters, the general public, and agency administrators. After the waterfowl season, managers must accommodate raccoon hunters, deer hunters, turkey hunters, bikers, hikers, fishermen, birders, nut gatherers, and planners needing data—all sorts of people. In addition, they have had to deal with food plots, co-op farmers, worn-out equipment, and piles of paperwork.

If the managers are the minds and hearts of their given districts, the wildlife technicians are the backbones and strong spirits that hold them up. Wildlife technicians often work alone in the hinterlands, where their sheer love of the outdoors allows them to be comfortable with mosquitoes, swamp water, mire, and unpredictable weather. These conditions only make them more determined to get their jobs done.

For waterfowl managers and technicians, the need to be near wetlands is visceral. They cannot help themselves—it is just there. Their vocations and avocations are indistinguishable from each other. When they are not on the job, they are hunting or fishing or birding or simply walking with their families through wetlands. To those with this sort of personal makeup, the problems that come with lack of funds, difficult weather, politics, discontented hunters, vandals, and inherited land unsuitable for their purposes can be frustrating. Yet, as these agents deal with ever-increasing demands, make do with small (though dedicated) crews, and tackle jobs with unpredictable outcomes, they accept it all as a way of life. In fact, they expect it. As a supervisor of these hardy souls, I came to know them well. It has been my good fortune that they have been people of such high caliber.

The staff in my district has numbered less than two dozen men at any given time. Among them have been Paul Brown and David Sams, who managed Reelfoot Lake. Ralph Gray and later Larry Armstrong managed five wildlife areas along forty miles of the Obion River. Alan Peterson, who was succeeded by Carl Wirwa, managed six wildlife areas along the Lower Obion River and on the Forked Deer River some thirty-five miles upstream to Alamo. Then there were the wildlife technicians—Robert "Corky" Morgan, Joe Carter, Dennis Whitson, Leroy Donnell, Ricky Quick, Bobby Norris, Jim Mullikin, Jim Griffin, Teddy Hobbs, and others—men who had the exuberance and energy necessary to confront the countless challenges in the field.

Larry Armstrong, the TWRA wildlife manager in charge of wildlife projects on the Obion River, including Gooch WMA and the Black Swamp project. (Photo by Jim Johnson.)

The Wetland Managers and Their Work

Nothing upsets people like these more than the inability to accomplish their objectives. As we "duck heads" discovered, our administrators too often found our interests too expensive and too difficult to tackle, and so gave them a low priority. This lack of support stemmed from four reasons: (1) land was costly to acquire, develop, and maintain; (2) manpower was limited; (3) the state's waterfowl objectives for the Mississippi Flyway were met; and (4) there was an ingrained perception that wetland development was for ducks alone. Indeed, there was some merit in such thinking when the costs and available funds were considered. Funding was shouldered almost entirely by hunters and anglers, and this source was insufficient. It was not that others with occasion to use the wetlands were apathetic about the problems; it was simply that they had no organized way to contribute. They did not require hunting and fishing licenses. With such inadequate funding, the state ceased to acquire and develop wetlands for years after the 1960s purchases.

Of all the reasons behind the state's long neglect of wetlands, the fourth one I cite—the idea that wetlands are for ducks and duck hunters alone—has proven perhaps to be the most shortsighted and difficult to overcome. The price tag for developing and maintaining wetlands certainly seems exorbitant when the only measure of success is the number of wintering ducks. But to make this the overriding criterion is inept and wrong thinking—wetlands are not, and should not be, only for ducks and hunting. Indeed, the need for public recreation, as great as it is, may well be the least of the critical justifications for protecting and managing wetlands. As I have noted, wetlands involve entire river ecosystems, and the benefits from these to all people are endless. One way to look at it is this: if managers can say that their wetlands are doing well, they can also say that the rivers are fine and that the people are being well served. Coming to grips with this essential truth is the only way the various parties involved with the wetlands can find common cause. Where will we end up if we do not?

Toward a New Age of Wetland Management

Those of us in wildlife management had long depended on tradition to make sure that the areas under our supervision were ready for hunters each season. By the 1980s, however, I and those working under me in what we called the northwest territories, or the northwest district, began to understand that our traditional management strategies were a mistake. These methods were part of the same subdue-and-control mindset that also afflicted the Corps of Engineers and private developers. Any successes we enjoyed proved to be temporary. As we tried constantly to buck nature, nearly everything we did was counter to the inclination of the river.

This attitude was especially ill-timed in the early 1970s, when I first began to work for the state. At no time in history was the wildlife manager more needed to ensure the survival and use of native wetlands than in the 1970s and on into the 1980s. However, our good intentions through this period outran our slowly evolving notions of how best to do the job. We had temporary successes, which fooled both ourselves and the waterfowl hunters. Encouraged, say, by three successful years of hunting within a ten-year period, the sportsmen would tell us, "Don't change anything. It worked once; it will work again—sooner or later." This was foolish thinking, however, and getting beyond it would take time.

Ultimately, we found that we had to yield to a powerful tutor—the laws of nature. Sooner or later, that teacher demanded respect. The never-ending dynamics of the river finally humbled all of us.

The stage for eventual change was set in 1974, when the Game and Fish Commission was reorganized and became the Tennessee Wildlife Resources Agency. Along with this change came more responsibilities for wildlife biologists and managers and broader objectives for the agency. More emphasis was placed on collateral environmental factors that impacted wildlife and fisheries as opposed to basic farming or habitat manipulation and game-harvest inventories. We became more concerned with factors like water quality and pollution, urban and industrial expansion, and the impact and trends of changing social attitudes. We also placed less emphasis on individual species of fish and wildlife and more on a wide array of species. Most important, we emphasized the management of the ecosystems that supported all of those species. As a result, concepts about the management of streams and wetlands began to change.

The agency's new objectives affected the wildlife profession itself, which now required formal college degrees from accredited colleges and universities. As more institutions provided degrees in wildlife and fishery sciences, new graduates began to fill the ranks of the reorganized agency, including, by the 1980s, most of the waterfowl-management positions in my district. With this new blood came a new curiosity about the "real world" applications of these young sciences, none of which raised more interest than those involving the study and management of wetlands. There were few experts with the necessary practical experience to consult for help, however. Thus, a period of trial-and-error management began.

The esteemed title of "duck head" changed soon after 1985 with the broadening of the managers' interests and objectives. Our jobs now targeted a wide spectrum of wildlife depending on wetlands—waterfowl, shorebirds, wading birds, neotropical birds, amphibians, and many others. After that, waterfowl managers were generally known as wetland managers.

Like many before us, we had been more than a little cocksure; we thought we could manhandle nature if necessary. It had long been easy to think that way, but when we were faced with the disappearance of the wetlands and the failure, more often than not, of our usual management methods, we changed our philosophies. As supervisor of the district, I came up with a new sermon: "Our job is not to control nature but to be guided by it." This idea was not really a revelation. It was something we inherently knew but had often not heeded until it became clear that we had no other choice.

"Hot Crops" vs. Native Duck Food

The problem of providing duck food offers an instructive example of how we had to change our thinking. One of the key traditional methods we used to accommodate waterfowl on public lands was to have domestic food crops, particularly corn, grown in the duck-hunting areas. Managers had long considered corn a "hot crop," a substitute for historical mast foods such as acorns, which, like corn, were high in energy. Both foods were frequently preferred by waterfowl during extremely cold weather, but most of the oaks were gone and with them the acorns. While corn has a high fat content, which can be an advantage for ducks when natural sources are unavailable—a female mallard must have 125 to 150 grams of fat stored before she will pair for the breeding season—a straight corn diet is disastrous. A mallard will die in approximately fifty days if fed solely on corn. This grain has only one of thirteen vitamins and minerals waterfowl need, and it is almost entirely unknown in natural wetland environments.

Competent farmers know that corn cannot really be grown in the path of the river where heavy flooding may ruin the crops. And we managers knew, deep down, that our practices ran counter to common sense. Before our water-management strategies began to change, we were in a constant struggle to keep nature from doing what it preferred to do. The month before the hunting season is October, the dry season, and following traditional thinking, we would pump as hard as we could from nearly dry streams to get water onto the corn before the hunters and the waterfowl arrived. But the crops were rarely successful for more than three out of ten seasons. Usually, as soon as the water was pumped to cover the food crops, it would rain. The streams would overflow, and everything in sight would be flooded. After the floods subsided, we often found our pumps clogged with silt, debris, and mud, and our levees nearly washed away. All our time and meager funds were wasted.

Another thing we knew deep down was that ducks had survived for eons on natural foods, long before we started giving them corn. Eventually, we began looking to the natural foods called "moist-soil" plants—what many think of as weeds—as a substitute for corn. Making this change, however, was difficult; it was not easy to justify cultivating weeds to a duck hunter. Experimentation, nevertheless, would prove that we were right; ducks responded well to native weeds. If corn is "duck candy," moist-soil plants are highly nutritious "duck spinach." When they are plentiful, they can readily accommodate the nutritional needs of waterfowl populations.

Mallards need both fats and protein to molt, nest, and raise young during migration. Ninety-five percent of their feathers are proteins; when molting, the birds require nine grams of protein a day. While protein is abundant in the seeds of moist-soil plants, even higher quantities can be found in "bugs"—that is, insects and other invertebrates. As it happens, these invertebrates thrive in moist-soil plants, particularly in late winter and spring when they aid in the plants' decomposition. As many as 178 species of inverte-brates have been found in such crops, and in addition to being a good source of protein, bugs are a good source of lysine, which helps increase fat deposition.

"Watch the ducks," a wise wetland manager will say. "They will tell you what they need." We did watch them, and by the mid-1980s, aided by some new research on the advantages of native food sources for waterfowl, we did our own experiments and substi-tuted natural food crops for hot crops, especially in those flood-prone areas where crops like corn consistently failed. Short-season crops, such as domestic millets, and native vegetation crops, such as grasses, smartweeds, and legumes, were planted as primary food sources for waterfowl, wading birds, and shorebirds.

This new food strategy often had dramatic results, which we saw at our WMAs and which, some years later, a couple of visits to a duck-hunting club operated by Harbert Alexander and Bill Dement of Jackson confirmed. Their facilities were on the South Fork of the Forked Deer River about eight miles south of Halls, Tennessee. They had spent long years planting corn for ducks and fighting the river just as we had. Their club was a duck hunter's dream. It included an all-weather road to a comfortable cabin with all the conveniences, a metal storage shed, and a covered boat dock twenty yards from the back steps of the cabin. Only one uncertainty kept them awake at night: the river could rise eight feet or so with hardly any warning. Sometimes it reached the top step of their cabin, and it could go higher during any season. They could tolerate the threat to their cabin but not to their corn crop.

Dement was curious about the diversified crops and low-level terraces that we were using on Black Bayou Refuge at Reelfoot Lake, the first waterfowl area where we began to test these new ideas. Here, we had avoided the high risk of planting corn on flood-prone land. Our philosophy was to plant corn only in places where a farmer would find it profitable; we reasoned that what worked for a farmer would work for us. We raised native moist-soil crops on the wetter sites where these plants were well adapted to short growing seasons. Every field had a buffer zone with shrubs, grasses, or hardwoods (mainly oaks). This type of management at Black Bayou Refuge provided a more complete and stable wetland than any resulting from our old methods, and it worked superbly for ducks. I suggested to Alexander and Dement that they abandon corn until more favorable land was available and do what we were doing at Reelfoot Lake.

Waterfowl on Black Bayou Refuge at sunset. This refuge was the first where wetlands managers changed their methods in order to provide habitat for a large variety of wildlife and a more stable and sustainable wetland for waterfowl. (Photo by Jim Johnson.)

The Wetland Managers and Their Work

Some months later, in March 2005, I agreed to meet them again at the club gate. Fearing that things might not have gone well, I arrived at the club gate about half an hour before the noon appointment to look over the area with field glasses. The river was not at flood stage, but residual puddles were scattered across the Forked Deer bottoms. To my surprise, a thousand mallards were feeding and frolicking in slash water only two hundred yards from the gate. This was half a mile from the club's main waterfowl unit. Obviously, things had not gone too badly for their late hunting season, although it had been rather dismal for most hunters.

Arriving on time, Alexander and Dement knew very well what I was about to witness. When we reached the clubhouse, which was still well removed from the main part of the club's land, we saw what must have been four thousand mallards feeding on natural wetland foods. Over the whole expanse of the club, probably twenty thousand ducks were feasting on the native food crops they had cultivated. In contrast, not a single duck was seen in the flooded cornfield on an adjacent club.

While the club had suffered through the regular duck season several months before (as most duck hunting clubs in the area had), it was a different story for the late hunt, when the ducks were migrating back north. Now, more ducks greeted them than there were hunters to hunt them. It was enough to convince these hunters that cornfields were not the only way—indeed, not the best way—to fill a marsh with ducks.

Another aspect of our new management plan included wetland cells for wading birds and shorebirds to accommodate their spring and fall migrations. These cells, which were usually constructed above the annual floodplain, were created using low-level terraces that followed, where practical, the existing contours of the land. Thus, there was little or no interference with the natural runoff or drainage around the cells and less likelihood of trapping water that would cause undesirable or unplanned ponds. These cells filled and dried more naturally, much like a meadow or an aged oxbow. Their design made them durable against strong currents and erosion caused by flooding. These areas provided a habitat not only for migrating birds but also for a wide spectrum of other wetland species, including frogs and otters. In addition, the ponds provided nesting and brooding grounds for resident waterfowl, and we noticed a significant increase in the production of resident or non-migrant mallards. In one season, for example, more than three hundred mallards were raised and matured to the flight stage at White Lake WMA (located west of Dyersburg near the Mississippi), and these were as wild as any ducks in the Mississippi Flyway. Indeed, by the early 1990s, we could see positive results in WMAs throughout the northwest district, including Black Bayou Refuge and new developments at Bogota WMA. By favoring the natural hydrology of the river ecosystem, terracing did not force floodwater onto neighboring land. In this design, the function of the river is unaffected, and although creating these structures might be initially more labor-intensive, better and more sustainable waterfowl areas were ultimately established.

River Corridors

Though small scale, the new practices we adopted in the wildlife areas had application to the management of entire river ecosystems. Diversity and protection of the wetland were the key principles. Eventually, warm-season prairie grasses and other native plants were added as borders to the wildlife management areas, because good field borders are essential to protect the wetland and frequently essential in the life cycles of wetland wildlife. In addition, these borders improve habitat diversity and provide a more ecologically complete and sustainable wetland. The intricate interrelationship of plant and animal communities does not stop at the edge of the wetland or the floodplain. In fact, the river ecosystem itself is incomplete without sufficient trees, grassland, and other native vegetation along its highland borders to complete the river corridor.

Such vegetated borders along the river corridor should be at least a quarter of a mile wide, though a mile would be far more suitable. These varied borders provide habitat for the "home range"—that is, the area required by individual wildlife species (such as wild turkey, deer, quail, furbearers, and turtles) during the seasonal cycles of the year. The home range of most animals living along the river frequently goes beyond the edge of the wetland and into the upland. If any of the requirements within the home range are missing during critical times, the animals may not survive. This consideration is nearly always lacking in the management of rivers. When little but open fields or other human developments exist beyond the river, wildlife species depending on rivers are severely limited in abundance and diversity.

Without those wide corridors of well-mixed native landscape, the river floodplain is, in a very real way, isolated along its reaches; it thus becomes a mere island remnant of an ecosystem. Ensuring that diversified corridors are in place is essential to the stewardship of a river. Wise stewardship fulfills the habitat needs of animal life in the river ecosystem and maximizes benefits to farmers and landowners adjacent to the river. Such river corridors result in fewer problems from flooding, better drainage of the upland, fewer conflicts with neighboring landowners, more profits from timberland, and abundant outdoor recreational opportunities. At the same time, all species of wildlife endemic to the area benefit.

It is not easy to summarize a wetland ecosystem in a few words, let alone understand how to manage it and then explain what is needed to the public. Thus, during the 1980s, as we wildlife agents in West Tennessee began to develop a decidedly different way of managing wetlands—one that saw free-flowing rivers as the key to solving persistent problems—we often found that we lacked the material means, as well as the expertise, for massive public communications. Gaining acceptance for new and unconventional ways of doing our jobs became a constant source of aggravation. But once we made a few effective innovations, we had no intention of looking back.

Chapter 6
The Obion–Forked Deer Wildlife
Management Areas

Among the many tests for our new management philosophy, one of the most daunting would involve the Tigrett Wildlife Management Area near Dyersburg. This WMA—first acquired in the 1960s (along with Gooch and Moss Island) and eventually expanded from about three thousand acres to some seven thousand acres—fairly well represents the swamps found throughout the Obion and the Forked Deer complex. Nearly 90 percent of Tigrett is under the influence of permanent standing water. If one wants to see how natural resources can be turned into a manufactured mess, this is the place to look. Fishing in Tigrett declined to a few diehard anglers by 1970. Waterfowl hunting followed this decline, and by 1985 prime hunting was largely confined to the upper extremes of the area, no more than a quarter of the acreage formerly hunted. As wildlife managers, we could do very little about it but scratch our heads.

The problem was the expanding artificial swamps. Many of these were created by private duck clubs in the mid-1950s. Soon afterwards, the oaks began to die. Waterfowl biologists did not try to deal with the problem until the state purchased the land a few years later, and the challenges only became more difficult over the next two decades, as many wildlife populations—including wading and shore birds, osprey, dragonflies, midge flies, muskrat, and mink—declined sharply or disappeared altogether. Aquatic vegetation such as water lilies, lizard tail, and button bush became common, along with frogs, bloodworms (chironomids), and a few other species. In stark contrast to healthy native swamps, with their abundant categories of plants and animals, the artificial swamps found in places such as Tigrett had a severely ailing ecology.

Duck Hunting at Tigrett and Elsewhere

Ducks certainly prefer more favorable habitat than artificial swamps. Around 1980, just as my colleague Gene Cobb was about to retire from the TWRA, I joined him and another colleague, Frank Zerfoss, in a last effort to hunt ducks at Tigrett. Gene was content to hunt off the railroad tracks that crossed the middle of the area. But Frank and I hacked our way through the buck brush and bramble to the middle of the swamp. After wading through knee-deep mud for half a mile, we reached our destination. A few ducks had pitched there. By the time we returned to the railroad tracks, we decided that the trip was hardly worth a couple of mallards. In fact, it was not worth much of anything insofar as we could determine.

Our frustration caused us to reflect on why Tigrett no longer supported ducks as some places farther upstream did. It was not rocket science; there was no duck food. From then on we went upstream to hunt. Although the living oaks were virtually gone, the swamps there would dry enough to grow native weeds, and weed seeds in water attract ducks. We also went to Anderson-Tully WMA near the Mississippi to hunt mallards among naturally flooded green trees. Here, acorns and the seeds of wild plants were still found in abundance. And we found success in the tall timbers of the Hatchie and at the edges of flooded fields in the Wolf. It simply required areas that followed nature's cycles of flooding and drying to produce good duck hunting.

Still, despite the deficiencies on the Obion and Forked Deer, other factors made that river system more favorable for ducks than the rivers to the south, like the Hatchie and the Wolf. Ducks had developed a habit of coming to the Obion and Forked Deer. The floodplains of this system have a wide, shallow gradient with crop fields at or near the margins. When the floodwaters rise, the shallow water covers vast acreages of the floodplain. At the edge are flooded crop fields, ideal for wintering mallards. In contrast, the southern rivers have hillier watersheds, the valleys are narrower, and fewer acres and fewer grain fields are found adjacent to them. Duck hunting is best on the Hatchie and Wolf during extremely cold winters when the Obion and Forked Deer are nearly frozen and ducks are pressed south by the harsh climatic conditions.

Dramatic changes were taking place throughout the region, but most waterfowl hunters gave them little thought. They simply went wherever ducks seemed to concentrate, not thinking much about the possibility that they might run out of places to hunt. Bottomland hardwood hunting had become a thing of the past for most of the Obion and Forked Deer as early as the mid-1950s. When the hardwoods died, hunters did as Frank Zerfoss and I did—they went upstream to the weed fields, to the edges of the swamps, or to other rivers.

To the surprise of many waterfowl managers, however, perhaps four times more duck hunting occurs in the Obion and Forked Deer today than in the 1970s, resulting in larger duck harvests. The reason is that hunters learned to build their blinds away from the annual floodplain to avoid the artificial swamps and lack of green timber. Instead, they generally sought the higher ground away from the encroaching, rotting swamps. Once there, they built water-controlling levees and provided duck food. To some extent, this was a reasonable option, but still the hunters often found themselves too close to the flooding river. Nevertheless, there were successful years for many hunters—and this despite the fact that wintering populations of ducks, according to our waterfowl surveys, have fallen by more than 30 percent.

The limited success of hunting in levied hunting grounds, however, comes at a price the hunter may not realize. While "duck levees" in the floodplain sometimes result in good hunting for a few years, they often do so at the expense of the river and thus at the expense of other sportsmen and fishermen, the bottomland forest, and wildlife and fish habitat in general.

Waterfowl populations began to fall along the Obion and Forked Deer during the 1950s, and the declines became more pronounced after the 1960s. Carl Yelverton, a biologist with the Game and Fish Commission, saw it from the beginning. He noted that hunters killed an average of 19,373 ducks each year in the Tigrett Bottoms between 1958 and 1960.[1] In 2003, less than 5,000 ducks were harvested at Tigrett. While the numbers fluctuate over the years, this most recent year saw a decline of nearly 200 percent. The loss of habitat obviously had something to do with it.

Inspecting the Encroaching Swamp

Whatever other problems might occur within the Mississippi Flyway, vanishing habitat has been the chief difficulty affecting the waterfowl populations in West Tennessee. Managers noted the losses at Tigrett more than thirty years ago. In May 1974 my colleague Floyd "Speck" Hurt and I met to begin to assess the damage to the bottomland hardwoods on the south side of Tigrett Wildlife Management Area. The dank, rainy atmosphere, reeking of stagnant mud and water, seemed to fit the scene that day. The stench worsened as we waded through the water and debris, stirring muck with each laborious step. It was disturbing to think that we had had anything to do with creating this decadent swamp shunned by both animals and hunters. Yet, according to Speck, the extent

Summer ponding on a bottomland hardwood stand at Tigrett WMA. The standing water will kill these trees if the floodplain is not restored. (Photo by Jim Johnson.)

The Obion–Forked Deer WMAs

of the dying and dead hardwoods had grown by more than one hundred acres in the previous ten years. The bottomland hardwoods—oak, pecan, sugar maple, hackberry, and hickory trees—had died one patch at a time as the area became wetter. The oaks that had once stood tall here were practically gone. Some of the trees had lived seventy or eighty years, and many measured three feet in diameter breast high. The swamp that killed them now encroached on the edge of the floodplain, and it appeared that it would eventually take over the entire south side of the area.

As far as the eye could see was a monotonous expanse of stark, dead snags, bleached white from the elements. Nothing much was left but the tall skeletons and a thick tangle of buttonbush, vines, dying water maple, submerged trees, and anaerobic mud. The snags, which looked like a dense wall in our field binoculars, had been standing like this for five or six years, according to Speck. The swamp took up nearly five hundred acres, about half the size of some swamps further upstream. Stagnant swamps like this were found in nearly every mile of the Forked Deer and Obion rivers.

Speck Hurt had been a waterfowl manager with the Tennessee Game and Fish Commission since it had been formed in 1949. He had managed Tigrett and Gooch from their initial purchase by the state in the 1960s. Speck had worked twenty-three years in these rivers before I joined the Commission. Still, he could not explain the what had happened here.

"Ducks won't be coming here much longer," Speck said thoughtfully. "There's not much food left for the ducks without the acorns. They seemed to stop coming to these new swamps six or seven years ago. We've lost half of the hardwoods in this four-thousand-acre unit since I was first assigned to Tigrett. The acorn crops are down to nearly nothing; oaks, you know, are first to feel the effects of all this summer flooding. It makes no sense."

The levees that were used to develop the impounded water in the old duck clubs had eroded several years before Tigrett was purchased by the state, and the old control structures had rusted to useless condition. Beavers plugged all outlets. The old club lands were now part of the expanding and stagnant swamps. "We used to see ducks all over this bottom," observed Waldon Guinn, who owned the cotton gin at Friendship near Tigrett. "Now we don't see them any more."[2]

Like others, I had grown accustomed to thinking of artificial swamps as pretty much normal, and so it was hard to understand at first why the ducks, which were supposed to like wetlands, shunned these particular swamps. The older swamps—like the one on the north side of Tigrett, near Roellen—had mulefoot bonnets, water lilies, and red-winged blackbirds, all of which tended to fool one into thinking the lands were being well managed. It was only gradually that we began to understand how, when humans nullified the natural relationship between the wetlands and the native rivers, biological and hydrologic decline was the inevitable result.

Grasping this essential truth came in large part through comparing the wetlands at Tigrett with those at Reelfoot Lake, and, later, with those of the relatively natural Hatchie River. Together, these three systems provided much-needed insights. This was new to me since comparing wetlands had not been part of my university studies. But I came to see that, even with the wide differences among these areas, a common thread connecting them kept surfacing. Natural hydrology was the key; natural hydrology produced the best results. The wetlands found along the channelized segments of the rivers were a product of levee-building, and these were not the same as the oxbows along the Hatchie.

The Floodplains at Tigrett

Tigrett WMA begins at Highway 412 about a mile upstream from the Dyersburg city limits, where it straddles both sides of the Forked Deer River. The management area continues upstream for thirteen miles before it stops at the junction of the Middle Fork. In 1966 the WMA contained 3,298 acres in scattered tracts that did not include the entire floodplain along any river mile. More acreage was purchased over the years and was close to 7,000 acres in the mid-1990s. The original Forked Deer River, no longer a natural river, has been replaced by an artificial channel, first dug around 1916. Since then, the channel has been redug and maintained with much the same alignment. The old abandoned segments of the river now lie stagnant and are practically nonfunctional. Most of the old river parts lay barely noticed beneath the artificial swamps. This segment of the floodplain now stands as a classic example of a channelized river.

The straight-line channel on the Forked Deer River through Tigrett WMA. Note the path of the natural channel on the left; this part of the old river no longer functions and is now part of an artificial swamp. (Photo by Jim Johnson.)

The Obion–Forked Deer WMAs

A dredging machine at work on the Forked Deer River during the channelization project. (Photo by Jim Johnson.)

Channelization on the Forked Deer changed the natural hydrology of the floodplain, failed to reduce flooding, and contributed to the ruin of the original river. (Photo by Jim Johnson.)

One disturbing lesson of Tigrett is that artificial rivers like the Forked Deer are in a continuous state of flux, never reaching the equilibrium found in natural streams. The nearest that such a river comes to recovery is when it is completely flooded. By then the evidence of channelization is largely inundated, along with the other features in the floodplain. Highways, high levees, and the tops of surviving trees are practically the only obstructions visible during high water.

A levee such as a highway, which has only a few outlets, can act as a solid barrier across the floodplain when the outlets are plugged up or too small to accommodate the flow of water. The river may rise six feet higher behind a highway than it will downstream at the crest of a flood. During these events, the water is usually chocolate-brown, heavily loaded with sediments. The velocity of the stream slows as it approaches an obstruction, such as a highway or a levee, and tons of sediment accumulate behind these structures. In forested areas the older, well-established trees appear to sink from the accumulating sediments. The once-exposed root collars of sixty-year-old trees and tree stumps can be found seven feet or more beneath the surface of the accumulated sediments.

When six feet of water press against a cross-levee or roadbed, the energy potential of the water becomes enormous and is especially violent wherever it seeks an escape, such as through a culvert or under a bridge. The entire flow of the river is sometimes forced through a single opening. Where there are no outlets, the river tries to go around the levee, over it, or through it. Whatever the case, the river will not be stopped.

Logs, trees, and other debris frequently stack up against the upstream sides of bridges. The ground vibrates from the constrained power of the river. The human-built structure is pitted against nature and is often tested to its limits. Engineers have wrestled with uncertainty involving these forces and the structural designs needed to withstand them. Sometimes bridge piers are ripped out of their foundations, or they sink in unstable soils, collapsing the bridge or even an entire section of highway. Earthen levees are always at risk and frequently wash away or become badly damaged. It leaves one to wonder why highways are so often built like levees instead of on piers to provide openings across the floodplain.

Once a flood begins to recede, more destruction follows. The disorganized river zigzags in all directionss in search of old pathways, or another favorable escape route. Finding few, the rampaging river plows at earthen impediments, uproots trees, scatters debris, and dislocates anything that cannot resist its defiant force. Dislodged material is often consolidated into a mass, only to create more obstructions in the river. This chaos continues until the energy of the river is exhausted. Still, it does not find equilibrium.

Finally, the chaotic river lies at an uneasy rest, and the floodplain becomes quiet. With no discrete natural channels or any way to distribute sediments in a natural and orderly fashion, raw mud and debris are dumped in fields, the forests are distorted, and the floodplain is generally wrecked. Trapped water stands still here and there, creating stagnant pools or adding to existing artificial swamps and generally contributing to the degradation of the river. With the next flood, the cycle is repeated, just as it has done for more than a hundred years. This is the nature of constructed rivers.

Beavers: The Scapegoats

Surprisingly, the chaotic and damaging hydrological changes like those at Tigrett WMA are rarely blamed on the root cause: sediments and channelization. Instead, beavers are blamed. Indeed, beaver activity has contributed to the problem, but they are not the cause. A lot can be learned about the functions of a river by studying the habits of the beaver. In a natural floodplain, beaver activity follows that of the river. Beavers are a part of the natural river ecosystem and are rarely detrimental to it. Timber interests might complain about the gnawed trees, but nature does not. In artificial rivers, however, it is another story: beavers take advantage of floodplains in chaos and worsen the problem.

The sculpted and straight channels of a constructed river are unstable, with the waters relentlessly trying to escape from their confinement and slicing through natural drains. They are incapable of draining the land sufficiently. In natural streams, after floodwaters recede to a certain threshold, they immediately seek well-defined, natural drains, and— as if orchestrated on command—the floodplains drain. In a channelized river, drainage becomes a contest between the excavated channels and the remnants of the natural drains. Since human-built streams are dug contrary to topography and natural drainage, the end result is chaos. The floodplain cannot completely drain. The presence of a single ditch can disrupt the hydrology of the floodplain for long distances.

In natural streams, beaver dams are built to only partially block the stream—that is, the level of the dam does not go to the top of the bank. Thus, the dam acts also as a spillway. It is futile for the beaver to build a dam to bank-level. If it did, the stream would simply route out another channel and go around the ends of the dam (always dictated by the landscape topography). Even people do not build dams to the level of the riverbanks without spillways. In a natural setting, a beaver builds its dam just high enough to hold water for the dry seasons, yet low enough for the stream to maintain its function. Consequently, the adjacent floodplain is not affected. Ponds created by beavers in a natural system are rarely more than a few acres in size. When they are larger, human interference should be suspected.

Carl Wirwa, a TWRA wildlife manager, walks across a beaver dam that winds across the flood plain for a half-mile or more. (Photo by Jim Johnson.)

At a certain threshold, receding floodwaters must make a choice about which stream they will flow into—the unnatural ditch or the natural drain. Whenever possible, the natural drains will be sought. Since ditches are rarely designed to follow the contours or topography of natural drainage, they usually interfere with it. The straight path of the ditch passes through all gradients—the lows and highs in the topography and the natural drainage. The result is hydrologic confusion.

The beaver has a greedy streak; it takes advantage of our mistakes, constructing ponds on every acre of a floodplain possible. Here, the animal's natural talents as a dam builder are exercised to the fullest extent. In the case of an excavated ditch, the beaver makes an exception to its usual dam-building techniques. Instead of partially blocking the ditch as it would a natural stream, the beaver builds the dam all the way to the top bank. When the ditch fills, the overflow fails to go around the dam as it normally would but instead spreads across the floodplain, seeking the lowest ground and any natural outlets. With the water thus escaping, the beaver assesses its progress. Its incentive is to quickly build more dams to catch the overflow.

After a certain point, floodwater moving across a wide floodplain ceases to rise and thus decreases in velocity. Here the beaver seizes the moment: it is no problem for the animal to find a strategic location to push up a little mud, along with leaves and debris, and build a low-level dam across the floodplain to stop the escaping overflow. Sometimes these low dams can wind across the floodplain for a mile or more before tying to higher ground. In the event that a dam is not high enough to satisfy the beaver, it increases the height and thus the size and depth of the pond.

When the problems caused by flooding and beaver activity (a natural process performed in an unnatural circumstance) become full-blown, the entire floodplain could end up under permanent water, with some ponds stretching for hundreds of acres. Whether beavers or humans build the artificial swamps, the results are much the same: a poorly designed pond that is usually in the wrong place, often where the landowner or wildlife manager desires living trees. These ponds come at the expense of plants, people, and animals, including, possibly, the beavers. Unrestricted and unable to reach equilibrium with their environments, animal populations tend to rise to their limits and crash; sometimes they are extirpated. Thus, a study of the beaver might be all we need to understand the management and ecology of rivers.

The exploits of beavers are obviously not the main reason why our wildlife areas and rivers are in trouble. Humans deserve that dubious credit.

Management at Gooch WMA

Carl Yelverton, the waterfowl biologist in charge during the early 1960s, surveyed most of Tigrett and made plans for its future management. As tradition would have it, his plans included levees, pumps, and other water-control infrastructure. Yelverton's proposals, however, were not implemented at Tigrett, since the managers were preoccupied with a similar development at the newly created Gooch WMA, a few miles north on the Obion River.

Public waterfowl hunting areas were a high priority in the early 1960s, and developing new areas like Gooch and Tigrett gained considerable favor within the Game and Fish Commission and from waterfowl hunters. The model plan at Gooch promised a new day for hunters, thanks to C. M. Gooch, who owned the largest portion of the original three-thousand-acre tract of land. An avid duck hunter, Gooch stipulated that the land be developed for waterfowl hunting. A thousand acres of the land were set aside for the model area, known simply as Gooch Unit A. Initial enthusiasm generated by the model plan put the development of Gooch ahead of the Tigrett plan, which would not be developed until some years later. But the Gooch plan, like the Tigrett plan and those of private developments, was flawed by underestimating the power and needs of the river, and its managers would suffer dearly from this oversight.

Gooch WMA eventually became a seven-thousand-acre waterfowl area (coincidentally, about the same size that Tigrett is today). It lies along the Obion River, where its downstream boundaries start at old Highway 51 between the towns of Trimble and Obion. The waterfowl area then continues upstream six or seven miles to the ICC Railroad, where the North Fork breaks away toward Kentucky. The subunits of Gooch are levied water impoundments that lie like contiguous boxes along the river, mainly on the south side. Some are three-sided units in which one boundary is determined by the high ground in the hills on the south side of the area. Floodwaters have nowhere to flow on this side of the river, except through some metal culverts.

Unit E on the upstream end of this series of units lies on the north side of the river between the main channel and the forks of the North Fork. Thus, the developments at Gooch have effectively inhibited the flow of floodwater on both sides of the artificial main channel. The only place for the river to flow freely is in the main channel itself. Levees bear the brunt of the river's force, and many suffer severe damage or collapse with each heavy rain. When the flood is over, the levees prevent, or interfere with, floodwater flowing back to the main channel. In situations like these, where water-control outlets are inadequate or unavailable, little interchange of water is possible from the floodplain to the main channel at normal stages of the river; thus, the water remains trapped behind the levees. If the water stands through the growing season, trees are likely to suffer severe stress or die.

Boat channels at Gooch were constructed across the cells to provide hunters access to blinds. The state built thirty-six waterfowl hunting blinds for the convenience of the hunters. Boats, cushions, and decoys were provided for the lucky hunters who drew lots for access to the blinds at a daily drawing. These pampered sportsmen could practically go duck hunting in their bedroom slippers, requiring nothing but a gun and plenty of shells, coffee, and food. During those early days, the state wildlife agency spared very little to spoil the hunters.

Gooch was opened to hunting in 1965 and became famous for its successful duck harvests. By the mid-1970s, however, the grand hunting area declined. It became too expensive for the Game and Fish Commission to staff the area with a manager and six or seven wildlife technicians. Half of the work force were transferred to other demanding assignments. The change left Gooch with a significant manpower deficit, another common result of shortsighted thinking. Those who remained were stretched beyond their capabilities, and that situation has changed little since. The convenient accommodations for hunters have long since been eliminated. The levees are still there—but only through the determination of the area manager. During flood conditions, the water has no place to go but over or through the levees. It does both. The levees erode nearly every year, and the maintenance costs rarely settle within the limits of the area manager's budget.

A large diesel-powered pump pipes water from the river to the internal units in a valiant effort to control floodwaters and maintain equilibrium. The pump is in constant jeopardy from flooding since the equipment is mounted on the levee of the main channel. Turbulent currents from high water regularly undercut the earthen foundation of the pump unit, and its intake pipe clogs with mud. These conditions keep the manager on high alert during fall and winter, when the river floods and the units are being filled. When the dry season arrives, the manager finds that sediments from the muddy floodwaters have filled the boat ditches; indeed, excessive amounts of sediment are deposited throughout the area. Water-control structures are often washed out of the levees. All facilities in the units remain targets for the raging floodwaters, and, for the most part, these structures are in constant need of repair.

Not unexpectedly, minor rains on top of the already-saturated floodplain can make it too wet for cultivating or planting food crops, or, if crops can be grown, they are often prematurely destroyed by floods. About 70 percent of the time, the crops fail. Even the successful crops provide minimal benefits to waterfowl. The units hold water up to five feet, and this is too deep for most feeding waterfowl: dabbling ducks, such as mallards, cannot reach food beyond a water depth of eighteen inches. Thus, large quantities of food, mostly native crops, are frequently unavailable to the ducks.

Approximately 25 percent of the model units at Gooch WMA still contain living oaks that provide acorns for waterfowl. Even so, these trees show stress from prolonged summer flooding and an excessive build-up of sediments on their roots, and the trees continue to die. Only through intensive efforts have the bottomland trees at Gooch been salvaged from large-scale destruction. Unfortunately, without floodplain restoration, the trees will all surely die, and the battle will be lost. (Sadly, hundreds of acres of bottomland hardwoods elsewhere in the Obion–Forked Deer floodplains have fared even worse than these trees.)

Unintentionally, we made a mess of the Gooch WMA. Dedicated to the conservation and management of wetlands, we had, over the long term, done exactly the opposite, albeit unknowingly. Had the project been developed out of the annual floodplain on higher ground, the management of Gooch would have been a success story. Better yet, it would not have interfered with the river's survival. But this alternative simply did not occur to land managers at the time; even if it had, it represented such a radical change from traditional management that it would have been rejected. Many of my colleagues and I look back today and wonder how much we could have accomplished had we known then what we know now.

None of these failures kept us from trying to solve the problems. Following our trail of disappointments as we struggled to change the prevailing subdue-and-control mindset will give the reader some idea about what the future holds for river wetlands and for the citizens of Tennessee. For the waterfowl/wetland managers in the northwest district, management areas like Gooch (and, as we shall see, Tigrett) were constant reminders of how easy it is be trapped by thinking that ingenuity and determination are enough to subdue a river to our will. They also stand as testimony that admitting our mistakes, accepting the rules for managing nature, and holding allegiance to good land stewardship are not enough to turn the tide.

Chapter 7

The West Tennessee Tributaries Project

Throughout the mid-1960s, the smell of raw, fresh dirt permeated the summer air a few miles north of the town of Obion. A giant crane leaned over the Obion River canal like a five-story praying mantis. Its buck-toothed bucket was large enough to hold a two-bedroom house. When it came to rest on the bank of the canal, a straight, open canyon ran through the floodplain. The canal was deep and wide enough to contain five freight trains side by side. The crane was finished with this section of channelization work, but it still had 170 miles of work to do on the Obion–Forked Deer river project. This massive public works project—a cooperative venture between the state and the U.S. Army Corps of Engineers—was known as the West Tennessee Tributaries Project (WTTP).

As we have seen, rerouting the river had started nearly half a century earlier when private landowners and local drainage districts became disenchanted with the slowness of the river to drain; channelization had been in progress ever since. The more the landowners worked, the worse the flooding became. Channelization was the only method they knew to reduce the flooding, but as direct and simple in concept as it was, it did not work. There is no record that anyone thought about addressing the real sources of the problem: soil erosion and too much encroachment on the floodplain. Since sweat and muscle had overcome most of their other problems, they saw no reason to change strategies. In this case, however, they were not applied on a sufficient scale to control a river. The expense, landowners claimed, held them back, and so they sought relief from the state, which turned to the federal government.

The corps agreed to come to the state's aid. All it needed was money and the congressional authority to spend it. U.S. Representative Jere Cooper of Dyersburg wanted to help. He had already sponsored the Flood Control Act of 1948 in Congress, and it cleared the way for the West Tennessee Tributaries Project. With state sponsorship, the corps met all of the requirements for starting up the WTTP in 1960. The corps's work was confined only to the main channels of the river, but the Soil Conservation Service soon came along with projects to dredge the small tributaries. With this, control of the entire river—the main channel and all of its branches—would be government-sponsored. The aim was to realign the channels and flush away the troublesome sediments that plagued the river and thus the flooding that plagued the people.

In 1951, writing in his *Sports Afield* editorial column, Michael Hudoba described Congressman Cooper's original intentions this way: "A new approach to the flood control problem is proposed by Rep. Jere Cooper (D-Tenn.) in his bill HR-1534. . . . Rep. Cooper would restore certain bottomlands for which levees are impractical and which restrict the flood channel back to the river. He would have the government acquire absolute ownership of these lands and they would be placed in Federal forest reserve to remain natural storage basins for periodic floodwaters."[1]

Obtaining public ownership of the entire river corridor was a novel idea, and it promised to heal most of the watershed: the forest could be restored, thus slowing the force of runoff and flooding; fish and wildlife and outdoor recreation would be available to the public; and the squabbling over who had ownership would be settled. It could result in the lowest-cost, most effective flood project the state had undertaken. The benefits could be enormous. Congressman Cooper had the insight to realize, a decade before the corps's work began, that the river needed to be free to flow—free from anything that would harm it and the public's interest. The congressman, who died in 1957, would undoubtedly be horrified by the warped outcome of his legislation.

The Corps Goes to Work

When the Corps of Engineers was enlisted for the project, it had already decided how rivers should be managed, and who would argue with the corps's success on the Mississippi? Looking on as the corps set to work, the state sat dumbfounded, but it was relieved to see the burden lifted from its shoulders. Conservationists, sportsmen, natural resource managers—indeed, practically anyone who was watching—could hardly imagine that they were about to lose their rivers. Farmers and landowners were exuberant that the plague of flooding was being confronted and, they hoped, subdued. Construction on the WTTP began early in 1961, and by the end of the decade, the corps had completed forty miles of channel on the Obion River and 14.1 miles on the Forked Deer River.[2] It was a small undertaking for those who had mastered the mighty Mississippi, but for West Tennesseans, this 225-mile channelization project was one of the biggest things they had ever seen.

Agriculture was a large industry in the fourteen-county watershed of the Obion–Forked Deer, and the WTTP responded to the private landowners' desire to improve drainage through their area. In addition to reducing flooding, the WTTP promised "an estimated 95 percent reduction in acreage of freshwater swamps, lakes, and sloughs."[3] Three-quarters of the Obion–Forked Deer floodplains were devoted to some form of agricultural production, and by 1971 half of this land was in soybeans, a short-season crop. World markets for soybeans had opened up new opportunities during the 1960s. Landowners with property in and around river floodplains saw enormous promise ahead: a crop that brought an average of four dollars a bushel might produce as much as ten dollars a bushel. For these landowners the WTTP was a godsend. The fourteen-county area covered more than one-tenth of the state. Farmers here produced 55 percent of the state's total crop production, nearly 400,000 acres of it in the floodplain of the rivers. The area also harbored more than a third of the state's bottomland hardwoods—242,000 acres—prior to the 1960s.[4] Many of the landowners began clearing new ground before the WTTP channel reached their land. From sunrise to sunset, smoke rising from burning piles of trees and debris could be seen on any day along the rivers. The value of bottomland hardwood lumber could not compete with the potential short-term value of soybeans.

By 1971 most of the forest not already in swamps had been cleared along the Obion and Forked Deer rivers. More cropland meant the need for bigger and more efficient farm equipment. It was a process that fed on itself. As the cost of farm machinery rose, staying in business required even more land and better efficiency. Agriculture became big business, and smaller farmers were quickly outdistanced. But converting wetlands to cropland was an especially risky venture, and the promise of soybeans often proved greater than the reality. In this gambler's game, many who took the risk went broke. But many would not give up. Larger trackhoes and bigger tractors could move a lot of dirt, build levees, and dig ditches much faster than before, and justifying them required that they keep busy. And so, modifying wetlands itself became a big business, one that was finally and disastrously overdone.

Land clearing during the 1960s and 1970s, such as this example in the Upper Obion, exposed fragile loess soils that eroded downstream and ultimately filled the river channels. (Photo courtesy of TWRA.)

Timberland was already at an all-time low in the region during the 1960s. Land-clearing and artificial swamps, encouraged by channelization, only worsened the deficit. The corps blamed beavers, which, according to its own early environmental assessment, caused bottomland hardwoods to die at the rate of three thousand acres per year. The corps predicted that channelization would help control the beaver populations—a view not shared today by wetland managers, who have seen the animals thrive in the conditions produced by artificial rivers and swamps. The U.S. Fish and Wildlife Service predicted that with the completion of the WTTP, few trees would remain in the area by the end of the century. But for the soybean industry, it appeared to be a boon for the region.

The state Game and Fish Commission could also see benefits from the massive channelization project. The first tracts on Gooch WMA were purchased around the same time that work began on the WTTP. The corps could lend a hand in the construction of Gooch when it passed through the area. In fact, the corps had an earlier agreement with the commission to develop the entire Gooch project, but it later reneged as mitigation for the damage caused by the WTTP became an issue. Nevertheless, the commission gained some concessions: since excavated soil from the river channel would be stacked for seven miles along the new channel, it could be used as a levee for Gooch development. The budget-lean Game and Fish Commission found this to be a generous contribution, and because of it, the construction of Gooch made rapid progress.

Optimism, in fact, remained high among most citizens along the river. But as the project progressed, others began to wonder if the cost was worth it. Still others wondered how it would affect waterfowl hunting. Channel work was completed in 1971 on the North Fork, and the work continued upriver on the remaining three forks—the Middle Fork, the South Fork, and the Rutherford Fork. But after seventy-one miles of completed channel, the corps began to run into opposition. Some forty-two landowners petitioned the corps not to dig on their land, which lent momentum to the rising ruckus about the project and whether it should be reconsidered.

Conservationists, wildlife biologists, and duck hunters asked pointed questions about the consequences of the new canal. One of them was Calvin J. Barstow, a waterfowl biologist for the Tennessee Game and Fish Commission. He would argue in 1970 that the WTTP ditch, seventy feet wide at the bottom and forty feet to the top of the bank (estimated to be about 30 to 40 percent larger than the original channel), was not needed to improve drainage in the Obion and Forked Deer rivers. Barstow and others believed that removing debris and cleaning out the old channel above mile thirty-two on the Obion River and above mile twenty-one on the Forked Deer River were far better remedies than the WTTP ditch and far less damaging to wildlife than the existing plan.[5]

Barstow's arguments were, at best, ignored. In 1971 Barstow's report on the existing WTTP channel work indicated that channelization was the primary cause of land clearing on the work completed and that it was responsible for the loss of 60 percent of the hardwoods on the Obion River.[6] More landowners began voicing opposition; some refused to give up the rights to their property. Wildlife interest groups saw a vast acreage of forest and wildlife habitat (as much as ninety thousand acres if the WTTP were continued) being lost to the soybean monoculture.

Conservationists in general commented very little about the possible fate of the rivers during these early debates. It was new territory to them; like the wetland managers, they had a hard time getting a grip on something as daunting as the fate of a river. Progressive technology, it seemed, had taken the lead; natural resource conservation took a back seat. It would be years later before the warnings of Ernest F. Swift, the director of the National Wildlife Federation, would come to haunt us. As he wrote in 1967, "This progress of ours . . . instead of creating a better place within which we live, [is] creating a place where, ultimately, we cannot live at all."[7]

The J. Clark Akers Lawsuit

The Corps of Engineers continued digging the canal until it passed the J. Clark Akers farm above Gooch Unit E. Akers was a duck hunter, and he did not like what was happening. In addition to the waterfowl hunting damage on his farm, some of his favorite hunting grounds upstream had been affected. He anticipated a dismal future for waterfowl hunting along the Obion River. Time would show that he was correct; many of the wetlands adjacent to the river channel were drained. The floodplain dried up, and waterfowl hunting declined.

Akers and three of his Nashville duck-hunting associates (members of the Middle Fork Hunt Club and the Davy Crockett Hunt Club) took the corps to court in 1970. They had their complaints, but the corps had also ignored Barstow's report, as well as an equally damaging assessment in 1959 by the U.S. Fish and Wildlife Service. Both pointed out that a substantial loss of waterfowl land could be expected as a result of the WTTP. This placed the corps in violation of the Fish and Wildlife Coordination Act of 1958.

The Coordination Act was the least of the corps's problems. A new law was about to be tested—the National Environmental Policy Act (NEPA) of 1969. Its purpose was to protect the environment from overly progressive government projects. NEPA stipulated that all major federal projects with significant environmental consequences be required to submit an Environmental Impact Statement (EIS) and be subject to exhaustive evaluation. The corps believed it had addressed environmental impacts in the environmental assessment it had produced early on. The courts, however, saw that document as weak and inadequate by NEPA standards.

In *Akers v. Resor,* a civil court ordered an injunction against the WTTP in 1973, effectively shutting it down.[8] At that point, only 32 percent of the project had been completed, 140 miles short of the corps's goal. No one could have predicted that more thirty than years later, the court's decision would leave the rivers no better off. The WTTP left a long legacy of lawsuits, countersuits, and wide disagreement over how the rivers should be managed. Yet, Akers's lawsuit was a hallmark for stopping one of the most damaging civil works projects known in the Southeast. According to Chester McConnell of the Wildlife Management Institute, a $100 million boondoggle had been halted primarily by the good intentions of a single duck hunter.[9]

Akers gained many allies for his bold lawsuit. The National Wildlife Federation and the Tennessee Conservation League (TCL), a state affiliate of the federation, joined the suit. Akers gained high esteem from the environmental community in this landmark case. The purpose of the lawsuit, however, was not to stop channelization but to collect damages for what had been done and for what might be done with the completion of the WTTP. Compromise between the state and the litigants would mean that channelization was inevitable. A large "gift" of land to the state in the form of mitigation lands would lay this issue to rest.

Preparation of an EIS is no small undertaking. These reports can cost hundreds of thousands of dollars and require considerable time, research, and public input. The corps had to hunker down and prepare for a long, arduous process. But they had resourceful and capable workers to deal with the document, and in 1975, after about three years of work, it was sent back to the court. The court would not rule on this revision, however, until 1978, and when it did, it found the EIS inadequate.

The EIS review was itself a long process, and meetings were held by the corps with interested parties in Memphis, Atlanta, and Nashville throughout 1979–81 to resolve the ongoing problems. Although the corps claimed a favorable benefit-to-cost ratio for the project according to federal guidelines, the plaintiffs refused to back off and continued to challenge the defendants. Finally, in 1985 and 1986, agreements were reached through more court action. Akers and his allies did not settle for the satisfaction of only temporarily stopping the WTTP; they sued for damages. Ostensibly, this action was not only for Akers's benefit but also for other duck hunters. Attached to the final decision of the U.S. District Court was a consent order that required the federal government to mitigate wildlife habitat losses already incurred by the previous work and the damages expected to accrue upon completion of the project. Ironically, the project was explicitly designed to drain thousands of acres of wetlands.

Throughout the long process of debates, court hearings, and rulings, friends and enemies were won and lost on both sides. In the end, the litigant gained more parties to his suit. The Tennessee Wildlife Resources Agency (a defendant in the original suit), the Wildlife Management Institute, and a string of others gave written or tacit support for the lawsuit. Open support by state organizations for the suit against the WTTP made for a delicate situation because the state had sponsored the project. The governor's office undoubtedly took a dim view of TWRA, which readily, if discreetly, supported the lawsuit. In time, this support would come back to haunt the agency as it became caught up in a form of blackmail between the promise of mitigated land and the duty to do the right thing.

In the final arguments, the government offered 14,000 acres to mitigate the damages by the work already completed by the WTTP. The U.S. Fish and Wildlife Service assessed the value higher—closer to 52,000 acres (it rose to 78,400 in their 1983 assessment). The state considered the value to be around 44,425 acres when it was fully developed for waterfowl hunting. Akers's lawsuit proposed 220,000 acres, a figure probably closer to the value of the lost lands than any other estimate.[10] Finally, in 1985 and 1986, a compromise was reached and the final figure was 32,000 acres, plus landowner easements. This acreage, known as the mitigation lands, would became a center of ongoing controversy.

For Akers himself, the long process must have seemed like tending a vat of fine wine. As time passed, the land would be developed into a long series of public waterfowl hunting units for miles along the Obion and Forked Deer rivers. The TWRA and the duck hunters were exuberant. The mitigation land would be owned and managed by the state for public hunting. The ruling also included a means for managing the land. An addendum, known as the Flood Control/Conservation Plan, dictated the design, development, and management of all 32,000 acres. This plan (one that would eventually conflict sharply with the wetland management principles held by the managers in the TWRA's northwest district) laid out seven reaches, three on the Obion and four on the Forked Deer. The plan, however, had ignored, overlooked, or misjudged one important compromise: the corps could continue to dig its huge artificial channel, much as originally intended.

At the time, the TWRA had only Gooch WMA as a developed public waterfowl hunting area. Everyone appeared to be pleased with the results of this new project, and the design of the mitigation lands followed the model at Gooch WMA for development. What they did not know was that ecological failure at Gooch would surface in plain view some twenty years later.

The Flood Control/Conservation Plan (which resulted in the West Tennessee Tributaries Mitigation Lands Wildlife Management Plan of 1983) involved managing two kinds of ecological features: "bendways" and dead-timber areas. It was an expensive wetland project as well as an intimidating one for land managers. The issue was not whether they could do it but how well it would work. The twenty-one bendways were segments of the old river meanders that had been cut off and left in the floodplain when the new channel replaced the river. These features were designed to include twenty miles of new dikes (or levees) and 13.9 miles of refurbished levees, three weirs (dams with water outlets), three diesel pumps, and sixty-six water-control structures. In addition, seventy of the artificial swamps, or dead-timber areas, were to be held as large ponds. Both the bendways and the dead-timber areas were intended to promote flood retention to slow runoff and to facilitate waterfowl management and other benefits to wildlife. By all standards, the plan was a massive undertaking for TWRA. At the time, it seemed to us to be a glowing example of conservation progress, but our hopes were misplaced.

The bendways and dead-timber units were strung along 150 miles of the Obion and Forked Deer rivers in designated and prioritized "reaches." Their distribution was a logistical challenge for waterfowl managers. To be managed effectively, an area of this size would need a sizeable annual operations budget, perhaps as much as a million dollars. Two dozen additional managers and staff were needed to operate and maintain the new land, none of whom had been accounted for in the project design.

The bendway and swamp features were objects of considerable debate and controversy throughout the EIS review. In comments attached to the corps's EIS, Col. Forest Durrand, a former director of the Tennessee Game and Fish Commission, said the plan "would come as near as possible to allowing all interests to 'have their cake and eat it, too.'"[11] Chester McConnell of the Wildlife Management Institute and others wondered how these work-intensive units would be managed. Some of the bendways and swamps would fall on private lands outside of the boundaries of the mitigation lands. Would these lands also be purchased? It was an important consideration since, without enough land, water held behind levees can encroach on private property.

Chester McConnell, the field representative for the Wildlife Management Institute, led the charge repeatedly against the West Tennessee Tributaries Project, the development of Black Swamp, and other wasteful practices that affected the rivers. (Photo courtesy of Chester McConnell.)

Still others debated which state agency should manage the wetlands. It was important—just as important to wildlife managers as the question of who managed the rivers. The corps's choice was the Obion–Forked Deer Basin Authority, which had been created by the state legislature in 1972 to maintain the WTTP once the channelization work was finished. The issue was finally settled by an agreement between the U.S. government and the state on February 23, 1988; following the agreement, the state wildlife agency was selected as the wetlands manager. At the time I was the assistant regional manager for Region 1 (later I would become the supervisor of the northwest district where these projects were located). I had concerns about accepting the new obligations. The latest settlement had raised more questions than it answered. Since the plan allowed for surging water down a straight channel, I and others wondered whether such surging water would cause more flooding downstream. And why, we asked, had an effective erosion-control plan not been put in place before the channel work started? How could the forest be enhanced by drying the floodplain if landowners, at the same time, were prepared to clear these lands for agriculture? How could stream morphology be expected to remain stable? Concerning this last question, we noted that the channel was designed for the specific purpose of moving the heavy sands out of the stream to increase its capacity to carry water. How, then, was it possible for the finely textured light soils along the stream bank to withstand the scouring effects? Why had the corps not considered other, less damaging alternatives than channelization, such as managing the old river with a light touch and concentrating on obstruction removal? Why had the corps compiled a document so voluminous and complicated that no one could cope with it? It was clear that the Flood

Control/Conservation Plan was going to leave TWRA with more headaches and many details to work out.

The corps estimated the rising price tag of the WTTP to be $75 million in 1984. But the expense of this "progress" was relatively easy to justify: with such channelization projects, the costs were *expected* to be high. The price, however, would go even higher. Anyone with dredging equipment such as a dragline or a track hoe, it seemed, was in the business of channelization. Dredging machines were as common as farm tractors in this area. They dotted the floodplains like towering dinosaurs found in the Mesozoic swamps.

Channelization went as far as the headwaters of the rivers and beyond. Where the WTTP stopped, the Soil Conservation Service continued. The SCS, with all good intentions, sponsored small-stream channelization projects throughout the watershed for flood relief, but their work simply added to the problem. The SCS eventually backed off of its channelization practices because of increasing opposition. The benefit-to-cost ratio was too low to justify the projects for the SCS, but the corps had set a determined course. It is difficult to even slow down a project with so much momentum. It appeared to be only a matter of time before the court declared the corps's EIS sufficient, lifting the injunction against the WTTP. The only alternative for the plaintiffs, it appeared, was to go back to court. No supporter of the original Akers lawsuit had the stomach or money for that, except for one: Chester McConnell. As a field representative for the Wildlife Management Institute, he had been involved in lawsuits before, and going to court did not intimidate him. McConnell could see the end coming, and he was not pleased. Settlement of the EIS meant that channelization was here to stay, and for McConnell channelization was wrong—period. It

Giant dredging machines like this trackhoe were as common as farm tractors in the West Tennessee river bottoms during the 1960s. (Photo courtesy of TWRA.)

was enough reason for him to continue his fight. This formed the crux of a fundamental disagreement between McConnell and others involved in the Akers lawsuit.

"They can go ahead and settle it," McConnell said in March 1984, just as a resolution of the fourteen-year-old lawsuit seemed eminent, "but if I can do anything to prevent an unsound settlement, I will."[12] He left no doubt that he was prepared to sue. Within two years, the courts had lifted the injunction against the WTTP, and Akers and his supporters, including the Tennessee Conservation League and its executive director, Tony Campbell, were anxious to see differences with the corps settled so that the promise of thirty-two thousand acres of mitigation lands could be fulfilled. The corps attempted to restart channelization work on a segment of the South Fork of the Forked Deer and applied for a state water quality permit from the Tennessee Department of Conservation (now the Tennessee Department of Environment and Conservation, or TDEC). McConnell appealed to the department in his effort to block the permit, but in 1988 the permit was issued. According to McConnell, he had initially won over the TDC staff, but to his disgust, their support succumbed to political pressure.[13]

McConnell was not finished, however. He appealed the permit decision in the courts, spending five thousand dollars of his money, and although he lost this battle, the press coverage that the controversy generated proved a key factor in the denial of a subsequent permit application by the corps. This time, McConnell said, he "made it miserable for [the corps] by getting something like 80 names on a petition from landowners against this project." McConnell and James Priestley, a farmer opposed to the WTTP, took the petition to Washington, and some weeks later Tennessee congressman John Tanner "promised no more channelization."[14] The corps withdrew its latest application in May 1989, stating, "The Corps will not proceed with further construction or acquisition of related mitigation lands for the authorized project until the State of Tennessee provides water quality certification and necessary rights-of-way for construction of the project."[15] For the time being, the WTTP had been dealt a crippling blow.

McConnell abhorred the thought that the WTTP would continue to dig the "Big Ditch," and he opposed anything that might allow it to go forward. He and Akers, cohorts in the original lawsuit, by now had become bitter enemies because Akers seemed to have no concern about the devastating effects of channelization. Akers wanted one thing from all of this—duck hunting land—and the acquisition and development of the mitigation lands authorized in the decision over his lawsuit would take care of that. He had many supporters, including, in the end, the TWRA. Although the TWRA preferred having the mitigation land without channelization—that was the position I took, along with thousands of hunters and fishermen—the agency could not have it both ways. The option was to go with the Corps-Akers plan. But for now the WTTP was at a standstill, and so was the mitigation lands package in the plaintiffs' lawsuit. The corps had purchased only fourteen thousand acres at this point, and this fell far short of the expectations of the plaintiffs and the TWRA.

The ultimate solution was not in sight for anyone at this juncture. To the corps, it appeared to be the deathblow for the WTTP; for the conservationists, it was a victory over channelization. For Akers, his coveted mitigation lands seemed to have inched out of reach. The Akers lawsuit created a terrible impasse between the litigants and anyone else who did not agree with the result of the lawsuit. Akers refused to swallow this bitterness; I came to think he never would. He wanted the remaining acreage authorized in the WTTP consent order with or without the corps project.

One of the most disturbing aspects of the WTTP mess was how a lawsuit that appeared to fly the highest banners for the conservation of natural resources could, in the end, cause such divisiveness. Indeed, various state and federal agencies—TWRA, TDEC, the U.S. Army Corps of Engineers, the U.S. Geological Survey, the Soil Conservation Service, Wildlife Management Institute, Tennessee Conservation League, to name a few—all staked out different ground in the WTTP controversy. The suits and countersuits surrounding the WTTP tended to spoil relations among well-intentioned administrators at all these agencies, and highlighted a lack of coordination and authority in grassroots conservation principles that have jeopardized the management of some of the state's most precious natural resources—river wetlands.

The West Tennessee River Basin Authority

Back in the early 1970s, just as the legal wheels in the Akers lawsuit were beginning to grind, some questions arose: Who would manage the WTTP once it was finished? How would it be managed? Somehow the state had to settle on a competent agency with the authority and responsibility to manage West Tennessee rivers, and it would require a comprehensive plan endorsed by the state legislature. The plan, it was thought, should be developed by a board of directors directly involved in agriculture, conservation, and natural resource management. The board should set the guidelines for management and regulatory functions. The plan should include a mandate to manage all aspects of West Tennessee rivers as a sustained natural resource for all citizens of the state. The agency selected—or created—should be a clearinghouse for all projects affecting the functions and use of the river, with regulatory authority following the board's guidelines.

In 1972 the Tennessee General Assembly established the Obion–Forked Deer Basin Authority. The title reflected its first assignment and original purpose: to sponsor and maintain the U.S. Corps West Tennessee Tributaries Project. Its responsibilities centered on providing flood relief within the Obion and Forked Deer rivers. Like the WTTP itself, the new agency's highest priority was to relieve flooding complaints from landowners, mainly to protect croplands. The strategy was to remove debris and other obstruction from channels to drain and maintain the floodplain. Sediment control was emphasized. In one of its reports, issued about a decade after its initial creation, the authority also expressed an intent to "restore the natural integrity of the floodplain" and noted, "Permanent or natural cypress-gum lakes, oxbow lakes . . . shall not be arbitrarily drained."[16] It seemed impossible, however, to reconcile such lofty goals with maintenance of the work accomplished by the WTTP, which had caused so much damage in the first place.

Richard Swaim was the agency's first executive director. His background as a University of Tennessee farm extension agent qualified him to deal with the problems farmers and other landowners experienced with flooding along the rivers. Swaim was a native of Madison County and knew the hearts of the West Tennessee farmers. He was avid about this work. He dealt with farmer's problems on a daily basis and understood their needs. He had a special understanding and sympathy for the landowners who tried to survive in and around the Obion–Forked Deer river system. Swaim was also a duck hunter, and he could appreciate wet feet, hot coffee, and flocks of decoying mallards. He understood the effects of erosion and how they affected the river channels. He had seen the farmer's cropland become wetter each year and the crippling effects that this inundation had on farm families.

In his new role as director of the Obion–Forked Deer River Basin Authority, Swaim set out with his characteristic exuberance to do something about it. It was a monumental undertaking, and his philosophy for managing rivers sometimes conflicted with my own and that of others. Nevertheless, he and I debated these issues in private, which engendered mutual respect for our differences.

A central problem was the lack of a comprehensive plan or vision for the rivers. What planning did occur was often piecemeal and inadequate. Channelization had been conducted according to the old subdue-and-control mindset throughout the watershed since the early 1900s with no continuity of purpose, or even the means to see it through. The Soil Conservation Service and the agricultural community literally lost ground through their efforts. Stream channels continued to fill with sediment and became progressively worse. Although the SCS required the representation of diverse interests on the workgroups that were assembled for its watershed projects, the vision of each group was highly constrained. I served on the advisory teams for several of these projects and found that because our comments were limited to the project at hand, the effects of the project on the environment beyond the job site were not contemplated. When inquiries were made, for example, about the effects of a tributary plan on a receiving stream, the standard answer was, "We cannot comment on projects outside of the authorized project area." Thus, when Dyer Creek in Madison County was channelized, the project's contributions of sand to the Middle Fork of the Forked Deer River were not even considered.

The Obion–Forked Deer Basin Authority devised a "comprehensive development plan" in 1983 for its work on the rivers, but it failed to address the sustained management of the waterways as natural resources. For years after its creation, the authority's methods of work often conflicted with the interests of the conservation community. Finally, the Tennessee legislature decided to change the mission and name of the organization in the mid-1990s. Rechristened the West Tennessee River Basin Authority, it ceased to be an independent agency and was placed under the Department of Environment and Conservation. Here is the authority's own account of its history and its need for a revised philosophy:

Throughout the 1970s and 80s, litigation over channelization and river maintenance techniques caused work efforts on the river system to be very sporadic. Lawsuits were finally settled in a cumbersome and restrictive Federal Court Order [the Agreed Order], signed in 1985. Additionally, federal and state environmental laws significantly affected river management policy [TDEC had refused to issue the required 401 Water Quality Certification permits for large-scale channelization projects] . . . and by 1987 shrinking federal and state dollars and lingering controversy caused state and local officials to realize that channelization and current management techniques were no longer practical or effective.

By 1994, legal restrictions, and other contentious issues surrounding the agency left the Obion–Forked Deer Basin Authority with limited resources, and a need for new legislative direction. Hence, the agency was reorganized by a bill passed by the General Assembly in 1996, and became known as the West Tennessee River Basin Authority. . . . The agency was administratively attached to the Department of Environment and Conservation. Additionally, the Hatchie River Basin was added to the Obion–Forked Deer River Basin as the agency's new area of responsibility, now covering over 8,000 square miles in all or part of twenty counties in West Tennessee. In 2003, the West-8 Area added over 2,000 square miles, with three new counties and parts of five existing counties along the Tennessee River.[17]

A new and refreshing opportunity had arrived for West Tennessee rivers. With it, a new mission statement for the authority was formally announced. The overall responsibility, under this new philosophy, was to "preserve the natural flow and function of the Hatchie, Obion and Forked Deer River Basins through environmentally sensitive stream maintenance practices." More specifically, the new objectives were (1) to restore, where practical, natural stream and floodplain dynamics; (2) to maintain or stabilize the function of altered streams and rivers; and (3) to provide regional and local leadership for the conservation and sustainable utilization of these river basins.[18]

New leadership accompanied the new mission. David Salyers was recruited as the executive director following the retirement of Richard Swaim. As a geological engineer whose philosophy and skills fit the new mission of the agency like a tailored suit, Salyers was well qualified for the position. However, a lack of funding, as well as a lack of enthusiastic support from the state legislature, slowed the potential of the Basin Authority, frustrating Salyers and all of those supporting his work.

Under the original objectives for the Obion–Forked Deer River Basin Authority, TWRA's proposals for restoring river wetlands often conflicted with those of the authority's leadership throughout the 1980s. The differences of opinions were largely over how dry the authority's plan and methods would make the floodplain. If the area became too dry, it would change the type of forest and most other vegetation in the bottoms—modifications that directly conflicted with the objectives of state and federal natural resource agencies. In my own view (and that of most of my colleagues in the resource agencies), natural drains could solve the problem; arbitrary ditches could not. Too little or too much water at the wrong time would cause the ecology of the river bottoms to change from its native historical norm. Restoring natural drains wherever possible in order to encourage and sustain natural flood regimes would, in this strategy, assure equitable drainage. Human developments within the floodplain should not be done where they compromise the ordinary functions of native streams, especially the rivers. In other words, any human changes in the historical functions of native streams should benefit the stream as well as humans. This is the only way, I and many others believe, that native streams can be practically or reasonably sustained for the greatest benefit to people.

Under this philosophy, the process of recovery should include sediment abatement, followed by restoration of the river channels in conformity with the preferred course of the natural stream. Impermeable surfaces, such as rooftops, concrete highways, and asphalt surfaces, which had long hindered the storage capacity of a large portion of the watershed should be phased out where possible. The Corps of Engineers, which had not been at all shy about modifying 225 miles of river, now has a perfect opportunity to show its best face and undo much of the damage previously done. Modifying the landscape to increase storage capacity, such as that provided in sediment basins, offers one possible solution to the problems of increased runoff and inordinate flooding. The West Tennessee Basin Authority's watershed flood-control lakes, some already constructed, appear to be one way to create a sponge-like storage effect that native watersheds provided with natural features. In a word, engineering and technology can now be effectively used to solve some of the artificial impediments that have been detrimental to successful stream management.

This is the philosophy of management that seemed to be unfolding in the Basin Authority's new mission statement. The agency's legal authority, however, still needs upgrading. The legislature has so far failed to face down local attitudes of "no government interference." Like it or not, however, rivers are not isolated to local concerns but involve those of the entire state. Lawmakers have not given the Basin Authority the means and enforcement capability it needs to meet its responsibilities and implement its new objectives.[19]

The state now has an agency in place with the right mission to turn around some seven decades of mismanagement for West Tennessee rivers. The Basin Authority—if it can keep its philosophy of management intact, if it does not acquiesce to unbalanced vested interests and unwarranted political pressures contrary to its principles—remains the logical choice to help us all regain an enormous natural resource, to help resolve a flooding plague to farmers and other landowners, and to obviate an inordinate tax burden to the state. All it needs is the financial support of the state legislature, a resolution to settle the WTTP, the confidence of the public, and the grit to face pessimistic bystanders. One day, Tennesseans will be immensely grateful to have this agency.

Chapter 8

Reelfoot Lake

"I see so much beauty in Reelfoot that sometimes I find myself worshipping the creation instead of the Creator," writes Lexie Leonard in his book *Reelfoot Lake Treasures* (1991). This native boat pilot for local state parks believes he can never say enough to express his deep love for this extraordinary wetland: "When I am not out there, I want to be, or I am thinking about it." He adds, "There's enough left to write another [book] and not repeat a thing I said in the first one."[1]

There are many like Lexie Leonard who feel that something would be wanting in their lives without the experience of native wetlands—the natural lakes, swamps, and lush marshlands filled with the sights and sounds of wild things. Being within sight of a place like Reelfoot Lake is often enough for a satisfying and a happy attitude toward life. Those who have truly experienced this area see it as far more than a swath of natural history. I should know—I am one. However, Reelfoot is not without its problems, and it became a learning ground for us. It was here that the guiding principles for managing West Tennessee wetlands were developed.

A Snapshot of a Unique Resource

As one travels east from the Mississippi levee in the northwest corner of Lake County, the delta spreads out flat and nearly treeless. From the levee, the traveler passes through some three miles of rich cropland—cotton, corn, and soybean fields. Some of these fields contain nearly a thousand acres. At the eastern edge of this open expanse, the traveler reaches Black Bayou Refuge at New Markham, the boundaries of public land. Here begins Reelfoot Lake, a twenty-six-thousand-acre wetland of immediate intrigue. Shaped like a sleeping fetus, it lies with its back almost against the eastern foothills of the Chickasaw Bluff.

Today, the upper extremes of this great wetland are mostly sloughs and marshland between ridges of hardwoods that have been harvested numerous times since its creation. The northeastern part of Reelfoot is mainly a federal refuge, which continues north for two or three miles into Kentucky. Southward, for some fifteen miles, the lake widens and

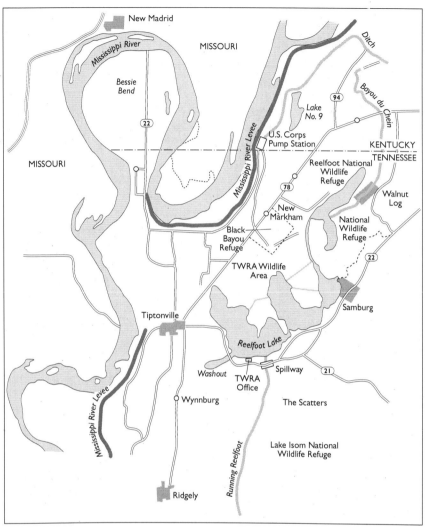

Reelfoot Lake.

constricts to form four major basins—Upper Blue Basin, Buck Basin, Middle Basin, and Lower Blue. The northeastern part of the basin, Upper Blue, is only a half-mile or so wide, but the southern basins, partially separated by peninsulas, fields of giant cutgrass, and shrub marsh, widen variously by as much as four miles.

A band of stately hardwoods generally border the northern two-thirds of the lake; the band becomes narrower, and frequently gives way to manicured backyards at the southerly end between Samburg and Tiptonville. Thus, the width of the hardwood band may vary from a few feet to more than a mile as the traveler takes the forty-mile trip around the lake. Varying layers of the forest canopy create vertical zones of habitat reaching upward more than 120 feet for different communities of birds, mammals, and other animal life.

Approaching the southeastern part of the lake, the oaks and hickory give way to green ash and, finally, cypress in the shallow water at the edge of the lake. Piloting a boat through this area, the visitor will see muskrats and beavers splashing and diving to escape the intrusion, while herons, egrets, and neotropical birds flush overhead as the visitor breaks through the cypress swamps. Proceeding through this zone, the traveler enters the open marsh with sparse trees and mixed colonies of giant cutgrass, purple-flowered tangles of swamp loosestrife, hibiscus, and rose hip growing on logs and stumps. The open lake follows.

The entire open lake—the epicenter of Reelfoot—is dotted with visible and subsurface stumps. Living cypress trees appear to thrive here and there across the lake's surface, as if they sprouted and grew in permanent water (more on this subject later). Continuing east, the traveler will encounter the forested tract known as Grassy Island—four thousand acres of hardwoods separated by marsh. Grassy Island stops at open fields of various widths that reach to the Chickasaw Bluffs.

Ecological changes occur in every direction along the journey. Intriguing nooks in various stages of ecological succession appear unexpectedly. Some are often so isolated that even nearby residents rarely see them. Giant cypress trees with trunks that four men together cannot reach around survive in isolated swamps. Cypress knees more than six feet high are found here. During the spring, one sees egret rookeries, bald eagle nests, cottonmouths, sora rails, least terns, fields of fragrant lotus blossoms, and perhaps a thousand more captivating features within a single day.

Trails more than a hundred years old wind endlessly through the marshes of Reelfoot, and these are still kept open by deer, bobcats, beavers, minks, other small animals, and boat traffic, not to mention maintenance by the lake managers. More than forty miles of maintained trails meander through the woody swamps and giant, cutgrass-laden marshes. Patches of cypress coves dot the open lake, which totals some fifteen thousand acres. These spots provide some of the best waterfowl hunting, as well as some of the best bluegill, largemouth bass, and crappie fishing in the state.

Reelfoot Lake is listed on the National Register of Natural Historical Areas, an indication of its fascinating cultural and natural history. The lake is a living laboratory for the study of wetlands, with attributes that are practically endless. One can only wonder why the state—and indeed the country—would not declare a high state of urgency to see not only that this natural treasure survive with its vitality intact but that it thrive culturally and economically to its full potential. One wonders, too, why we have not recognized that there can be a symbiotic relationship between people and a great natural resource like Reelfoot Lake. Such a relationship most certainly can be realized but only if the state's citizens are willing to do it. Such willingness can come with better knowledge of this historic wetland, and the best place to start is by understanding the history and basic concepts of the lake's biology.

From Natural Wonder to Exploited Preserve

Reelfoot Lake is an oxbow swamp—but only in part. It is also a large earthquake lake. This makes it unique, not only in Tennessee but in the world. It is an oxbow because it was created from a segment of the old Mississippi River. It was originally a marsh and cypress swamp (and in fact, a lot of it still is). Then, the entire thing was shaken up, sunk, and filled with water. This dramatic process created a natural wonder that would draw international attention.

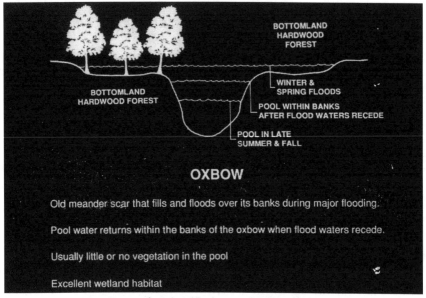

Diagram of an oxbow. (Illustration courtesy of the TWRA.)

Reelfoot as we know it today is young in geological terms—not quite two hundred years old. Perhaps hundreds of years before the present lake, another lake existed at the same site. This natural lake—an oxbow—was formed when the Mississippi River, yielding to the endless forces that changed it, adjusted itself by an instantaneous switch. This disturbance, which affected fifty thousand to sixty thousand square miles, left a conspicuous bend west of Reelfoot Lake—Bessy Bend. The switch stretched the river channel west a few miles, around a pendulum some twenty to thirty miles in length. It was a simple adjustment in the river, something rivers do from time to time. The switch helped to slow the river's velocity, relieving the forces upon it and helping it achieve equilibrium. The abandoned section—not yet known as Reelfoot Lake—aged as natural succession took its course. Eventually, the wetland evolved into a cypress swamp. Probably very few if any European settlers saw the beginning of Reelfoot—that is, while it was an oxbow. Perhaps someone saw it as an aging swamp, just before the earthquake. Indian artifacts indicate that Woodland Indians, however, frequently hunted and fished here sometime during this early period. Maybe as many as a thousand years passed between the switch to Bessy Bend and the occurrence of the most cataclysmic tectonic activity in North America—the New Madrid Earthquake of 1811–12. With a magnitude of more than 8.0 on the Richter scale, the quake sunk or re-created a large part of the aging oxbow and its outlying swamps. These swamps, according to some of the old maps, ran south from Kentucky about thirty miles to the Obion River. Only a fifteen-to-twenty-mile stretch of this swamp, from western Kentucky south into Tennessee, sank or was disturbed enough to form Reelfoot Lake. The disturbance of the earthquakes—there were a series of these quakes during the period—might have sunk the cypress swamp twelve feet or more. The earthquake probably influenced the creation of other lakes, such as Davy Crockett's Obion Lake, Open Lake, and Chisholm Lake in Tennessee, as well as St. Francis Lake in Arkansas, but none compare to the magnitude of changes that occurred at Reelfoot Lake.

Runoff and the waters of the Mississippi River filled Reelfoot over some period of time. The aging forest began a long process of dying (nearly two hundred years later, some of the cypress trees still survive in the open lake) after the lake flooded. During the first few years, the lake must have appeared as a flooded green forest. Most of the scattered water maples and oaks died within a year, since trees do not tolerate permanent flooding. Even water-tolerant cypress trees cannot survive indefinitely. We know of no other example where a natural swamp was reconfigured by an earthquake. The lessons wetland scientists can find here are probably unmatched by any place in the world—it is a unique spot for studying river wetlands.

Ironically, what might seem like the embryonic stage of Reelfoot's development—its earliest stage—was quite the reverse. Within mere months after it was created, Reelfoot Lake, as measured by human-user standards, had already reached its peak: It was in its most vibrant, resource-rich stage. Giant hardwoods climaxed the forest, and fish and wildlife filled the voids of this freshly created wetland ecosystem.

Although the fertility of natural lakes is typically high in alluvial valleys like those that surround Reelfoot, this lake had an unusual fertility from the time it was created; metric tons of organic matter from the litter and decay of the forest ended up on the lake bottom as a result of the earthquake. The scouring effects of the powerful Mississippi River currents were not enough to sweep away the rich forest soils and highly fertile organic matter—the dense, sunken forest managed to resist the force of the water. Thus, the lake became an instant nutrient-rich sink.

Fish, wildlife, and aquatic communities of fauna and flora must have flourished to the astonishment of those most familiar with oxbow lakes—even today the lake is more productive biologically than any lake in the state, perhaps the world. If the dynamics of Reelfoot had stopped here, decomposition of the rich organic matter undoubtedly would have consumed too much of the free oxygen for many fish and other aquatic life to survive. A massive die-off of these species certainly would have been expected. But there is no way to confirm whether or not a major fish die-off happened during this early history. If so, it was probably of minimal effect. This state probably allowed the lake to age in a fashion typical of large, natural, inland oxbow lakes—and all of it in concert with the dynamics of the flooding Mississippi River. Fish and aquatic life were stocked to optimum levels. The entire region from western Kentucky south to the Obion River flooded annually, and most of it, according to some maps, was swampland—sloughs, ridges with hardwoods, and drains.

The land was considered unhealthy and not worth much when the first settlers contemplated moving into the region.[2] As catastrophic as the New Madrid earthquakes may have seemed to the Reelfoot Lake region, they were like the great floods on the Mississippi—these dynamic changes were necessary to produce the magnificent creation we see today.

Within a decade or two after the earthquake, half or more of the cypress trees within the open lake were dead or dying. The tops of dead trees became conspicuous snags, bleached by the weather and mixed among the surviving green canopy. Old men in the early 1900s testified that some forty years after the earthquake, "there was a great deal more timber standing in [Reelfoot Lake] above the water" than there was at the time of their testimony.[3] By that time, most of the dead trees they remembered had been snapped off by strong winds at the surface of the lake. Thick sheets of ice pushed by the winds sheared off other snags at the same level, and many of the treetops sank to the bottom, adding even more organic matter to an already nutrient-rich lake. More have fallen since to exacerbate the accumulation of organic material and thus its fertility.

Within perhaps seventy-five years, the lake acquired the character of an aging swamp and appeared to be an open body of water. A hundred years after the earthquake of 1811–12, the state courts made note of this transition: "At present there is a government levee opposite the north end of the lake, which prevents the waters of the Mississippi at the ordinary tide from flowing into the lake. Before this levee was built, the river would flow into the lake once or twice a year, and thus raise the waters many feet. The surplus waters would remain in the lake until the late spring or summer following."[4] Reelfoot Lake had become one of the most, if not *the* most, productive freshwater natural resources in the state, or perhaps even the world. As the records of a 1913 lawsuit pitting the state against the West Tennessee Land Company would note:

> From 300 to 400 fishermen take fish from the waters of the lake daily and employed in this capacity something like 1,000 small boats, canoes, and bateaux. They take annually from 1,000,000 to 1,500,000 pounds of fish. Wild fowl are killed upon its waters every year in large quantities. Its waters are clear and pure, and as a breeding ground for fish, it is probably not excelled anywhere in the world. Before the building of the levee above mentioned the fish would have free access from the lake to the Mississippi [R]iver and from the river to the lake during high water. The public have used the lake and its fowling and fishing privileges at will for more than forty years.[5]

Waterfowl hunting has also been extraordinary at Reelfoot. According to a 1988 TWRA planning report, "One thousand ducks per day is the average estimate to be bagged by the sports and those who shoot for profit in on single day; including all kinds, twenty-eight hundred is the largest shoot that has been known in recent years. It is said, however, that five thousand were bagged in a single day about the year swan became a nonresident of Reelfoot Lake in 1873 . . . but they were burnt out with powder and lead until they have become unknown to this body of water in recent years."[6] My grandfather Wily Johnson, who was a young man in his prime at the dawn of the twentieth century, told me, "Ducks? Why, we rarely would shoot until we could kill more'n twenty—with one shot. Not like that today; these newcomers shoot a half-case of heavy-load shells to get one duck; shells're too expensive for this, you know. We'd bag more'n a hundred most days."

Such statistics and anecdotes give an idea of just how productive lakes produced by nature can be. But differing ideas over how such lands should be managed, who should manage them, and how human exploitation should be regulated can also be the source of rancorous, even violent disputes. This was certainly the case at Reelfoot.

Ownership Conflicts and the Night Riders

For years prior to the Civil War, the ownership of Reelfoot had not been an issue since the area's low, marshy land was deemed unsuitable for development. After 1865, however, lake ownership became a persistent problem. Farmers cleared some of the nearby land for crops, while others settled along its shoreline, pursuing their livelihoods through commercial fishing, trapping, or providing services for the sport fishermen and hunters who visited the lake.[7] By the 1880s, railroads connected southern lumber resources with northern marshes, and the lake was recognized as a valuable asset.

Despite various claims and legal disputes involving ownership of the lake throughout this period, local residents usually ignored these claims, hunting and fishing as they had always done. But the ambitions of one entrepreneur—James C. Harris—would eventually lead to bloodshed. Harris, a North Carolinian who came to Reelfoot in 1865, developed a plan to drain the lake in order to harvest timber and increase cropland. For twenty years, he bought up land in and around the lake and, in August 1899, announced his intention to excavate a ditch from the southern end of the "Washout" to the Mississippi River. Those whose way of life depended on the lake hired lawyers and argued that Reelfoot was public domain and that Harris had no right to drain it. A legal battle ensued, which culminated in a 1902 Tennessee Supreme Court decision that upheld an injunction halting Harris's plans to drain the lake. The court, however, also ruled that the lake was susceptible to private ownership. Harris died in 1903, but his son Judge inherited his holdings. While it is unclear whether Judge Harris meant to pursue his father's goal of draining the lake, he certainly wished to exploit it. In 1907 the younger Harris entered into a partnership with several others to form the West Tennessee Land Company. The company's claims to ownership and the restrictions it placed on local residents whose livelihood depended on the lake led to the rise of a militant band, the Night Riders, and violence against the company's representatives. Armed rebellion reached a climax on October 21, 1908, when the Night Riders kidnapped two attorneys, R. Z. Taylor and Quentin Rankin, who were members of the land company. Before the evening was over, the band of armed men had hanged, shot, and killed Rankin on the banks of the Bayou du Chein. Taylor escaped certain death by bolting into the bayou and the northern swamps, as the Night

Riders fired their rifles and shotguns after him. The local militia was called in to quell the civil unrest, and a long series of legal battles followed.[8]

Eight Night Riders were tried and convicted on murder charges, but this decision was reversed on technicalities by the Tennessee Supreme Court in June 1909. Meanwhile, the issue of lake ownership wound its way through the civil courts, and again the state's highest court had the final say, ruling in 1913 against the West Tennessee Land Company and allowing the state to condemn and purchase land in and around the lake. In 1914 the state came into complete possession of the lake for just over $36,000, including legal fees and court costs.[9]

The first state boundaries that defined and enclosed this public land were surveyed and recorded by H. M. Golden in 1915. The markers on Golden's line were practically destroyed before the survey was completed, and the line became difficult to maintain. This line—the ordinary low-water mark of the lake—was almost immediately disputed by local landowners because the lake levels continued to rise and fall, leaving a large strip of land between the high-water mark and the low-water mark in dispute.[10] The Tennessee Supreme Court recognized the rise and fall of Reelfoot as a natural phenomenon, but by 1917 the influence of the Mississippi River on Reelfoot was being curtailed by levees and other developments. These included a levee from Hickman, Kentucky, south along the banks of the Mississippi to high ground near Tiptonville, Tennessee, as well as a levee, spillway, and drainage canal at the lake itself. One goal of this construction was to hold the lake waters at a constant level so as to control and stabilize the state's boundary line, thus allowing it to be easily identified and maintained.[11] However, the new line was ignored by some: clubhouses were built, timber stolen, and crops were planted within the public boundaries. Legislation enacted in 1917 directed the state game warden to keep the waters of Reelfoot Lake at the same level in an effort to reinforce the state's authority over the lands.

It seemed, however, that neither the rulings of the Supreme Court nor action by the legislature could settle the disputes over boundaries. In 1925 the governor appointed the Reelfoot Lake Commission to resurvey Golden's line, since it was considered inferior, having less than one hundred lines and angles. The commission intended to establish a more precise and legal description of the state's boundary. Because of the encroachments since 1908, the boundary line at the ordinary low-water mark was compromised—the new line was the low-water mark minus one foot in elevation. But land acquisition became a nightmare. One hundred cases were litigated by the courts, according to the commission's final report on August 10, 1931: "We found lands held by unrecorded deeds, conflicting titles, and disputed lines in between claimants, conflicting, or disputed, titles through inheritance and other collateral issues."[12] This struggle with the boundary line and land purchases for the buffer zone continued until the commission was disbanded around 1931, mainly because of the governor's dissatisfaction and financial strains resulting from the Great Depression. The disputes are still in court as of 2005.

It seems strange today that the water levels of the lake were "managed" for the purpose of clarifying property boundaries. This subdue-and-control method became one more problem that ignored the needs of a natural resource. Today it contributes to the lake's stagnation, adversely affecting its vitality and management. Had the state realized what the full implications of stabilizing the lake levels were, it might have insisted that the levels be managed for more fluctuation, and the boundaries would probably have been set at a higher level. That way, there would be fewer reasons for anyone to complain about the ordinary rise and fall of the lake, and the lake would be in much better shape.

In order to control the water levels of Reelfoot and hold them at the level established by Reelfoot Lake Commission, a concrete structure with control gates was ordered and built. This meant closing all but one of the five outlets at the spillway site mentioned earlier. A new spillway was constructed at this site in 1931, which became a dam and part of Highway 22 connecting Tiptonville and Union City. The operation of the spillway became a duty of the manager of the eight-thousand-acre Reelfoot National Wildlife Refuge established by a 1941 lease agreement with the state. The cooperative agreement spelled out the joint duties of state and federal administrators for the state-owned lake and adjoining lands. It was the refuge manager's responsibility to regulate the water levels of the lake by manipulating the spillway gates. The flow of water through twenty openings at the top of the spillway is controlled by drop logs, collectively called gates. The drop logs consist of metal or wooden boards at the tops of the openings; they can be manipulated to increase or decrease the water levels of the lake by as much as eighteen inches. In addition, one gate installed in 1958, has the flowage capacity of all twenty drop log gates. This gate, operated by a crank and gears, opens from the bottom and has the potential to lower the lake by as much as 5.8 feet. Lowering the lake to the bottom of this structure, however, has proven unfeasible since sediments have accumulated in the outlet canal and underground springs, and seepage exceeds the outlet capacity of the gate as lake levels decrease.

While the intentions of the Reelfoot commission were honorable and often effective, the commission failed when it came to the biological management of Reelfoot Lake. There is no evidence that the commissioners had the assistance of natural resource managers, scientists, or anyone else concerned with the biology and management of the lake.

Social disputes and the modification of wetlands that have resulted from conflicts such as those exemplified by the Reelfoot Lake spillway are typical of the circumstances that create the flawed strategies for managing wetlands in West Tennessee. Sometimes, in ignorance, the public accepts policies and practices it would not permit otherwise. When this happens, projects such as the WTTP on the Obion–Forked Deer rivers and the spillway at Reelfoot are the results, and these are almost always detrimental to the natural resources in question.

Living in Harmony with the Lake

Not everyone living at Reelfoot Lake pitted themselves against the nature of the wetland; rather, they lived as a part of it. The people at Walnut Log, for example, tended to accept floods and low water. Until around 1960, this sleepy community was one of a handful whose sole existence depended on the continued vitality of Reelfoot. The residents' livelihood was jeopardized only by civilization's interference with the normal ecosystem of the lake. Until the late 1950s, a hundred or so inhabitants of the community were hunters, trappers, fishermen, and guides, and their families. I lived there from 1940 through most of the 1950s. My "backyard" ran from the back doorsteps west, south, and north through swamp, marsh, and the open lake. Nothing could be more satisfying for a boy than twenty-six thousand acres of natural wetlands for a backyard, but today it is practically a ghost community.

Located at the northeast end of the lake and straddling the Bayou du Chein, the community was, for the most part, self-sustaining. It had no mayors or town councils—just a few self-educated advisors. Its inhabitants did not depend upon Social Security or government food commodities for retirement before the 1950s; all they needed was the lake and help from younger and stronger backs. This was their retirement. They hardly felt the Great Depression. Financially secure sportsmen still needed the outdoors, and they continued to come for guide service and the recreational opportunities still plentiful at the lake. Guiding fees of twenty-five dollars to thirty dollars per sport, plus tips, depreciated very little, even during the 1930s. The inhabitants here needed little cropland: a garden was welcome, but all they really needed was the lake.

The only access to my parents' house from a dead-end road on the east bank of the bayou was by footbridge or boat. The land on our side of the bayou was about three feet lower than the bank on the other side. We lived *in* Reelfoot Lake, not *at* Reelfoot Lake. All of it was in the Mississippi River floodplain.

As a boy during the early 1950s, my interests focused on catching the evasive three-toed mink, or catching a glimpse of the fabled last panther, last otter, or last beaver. My grandfather had seen the last whitetail deer when he was about twenty; my grandmother the last panther. But no one could remember when the last wild turkey was seen. Civilization had already taken its toll on the wild game.

Becoming a breadwinner started at an early age for most of the young men at Walnut Log. At the age of twelve, as the oldest in a family of six children, I ceased to be a boy. By the end of that year I was accepted as a legitimate guide, commercial fisherman, and trapper.

With all of its conveniences, Walnut Log had a few things a mother could do without. High water was one. Flooding from the rise of the Mississippi River could be expected at any season, and I remember some of those events. Walnut Log Lodge, like several other buildings of the community's early days, was built on a raft of logs in anticipation of the floods. Eight-foot piers took the place of the log rafts before I could remember. On two occasions during the 1940s and '50s, the water was ankle deep when I stepped from my bed. Within a couple of weeks, however, the high water was nothing more than old news. Residents in Samburg, Gray's Camp, and other communities around the lake also experienced the effects of high water.

Onice Strader is one who remembered the flood of 1937. "It was about mid-January," she said during an interview in the spring of 2005. "I remember because it was about two weeks after I was married on January 6. Daddy had built the first store here, and it was later known as Gray's Store." "Miss Onice" lived across the lake a few houses north of the store

Onice Strader, a lifetime resident at Reelfoot Lake who lived through the 1937 flood on the Mississippi. (Photo by Jim Johnson.)

in the community known since as Gray's Camp. When the rains began in 1937, Gray's Camp received warning to move to higher ground, and that is what "Miss Onice" and her neighbors did.

"Well, Mr. B. L. Austin came by about then in his trucks," she continued. "He was a good man. He said he had a big sweet potato barn at South Fulton, and it was clean except for a few sweet potatoes in the loft. He gave us permission to use it, and we agreed. He loaded us up—all the women and children, groceries, beds . . . what we thought we needed for the stay. It had no lasting effect on the living tempo of the communities. We stayed a couple of weeks and came back home. Within a few days, the river fell; the land dried and everything was put back in order. Life went on as usual." That was the way it was with the people living around Reelfoot Lake.

There were some, however, who were glad when the Mississippi levee was finally built to its highest elevation. Most felt the same about the road/levee on the south end of the lake, since the Mississippi typically backed up some forty miles through the Obion River and Running Reelfoot.[13] The flood went nearly to the top of this levee. "They talked about watching houses floating down the [Mississippi] river, only their tops showing [in 1937]," Miss Onice recalled. The Mississippi rose to 59.5 feet at Cairo, 19.5 feet above flood stage.

Many reasoned that the Mississippi should flow through Reelfoot; the problem was how to do it once the development of the region had adapted to the security of the levees. I asked two longtime commercial fishermen at the lake what they thought about the

Mississippi River levees. "The lake needs this river water, but we could do without a '27 or a '37 flood," Ronnie Johnson said.

"What we are probably doing," said Frank Gooch, "is setting ourselves up for the 'Big One.' If we hold back all that water and the damn levee breaks, what do you think we'll be in for then? We made it OK before the levee. . . . There ought to be a better way."

But if one knows the risks and is willing to take them, what is there to complain about? The history of the communities around Reelfoot Lake seems to suggest that people can prepare for flooding and live quite well in certain floodplains without manhandling the river. But "lakers" like Miss Onice could be the exceptions rather than the rule. During the 1980s, the managers at Reelfoot Lake had to cope with diverging opinions about solving such problems as stagnation, declining fisheries and duck populations, and the ruin of surrounding croplands because of unnatural flooding. Common sense told us to determine the lake's current biological status, lay out the options, and ask for the people's support. How they would respond was the big question, but the answer, as we shall see in later chapters, would come soon enough. And it would not be pretty.

Bennett "Ras" Johnson shares a Bayou du Chein boat ride with his great-granddaughters, the latest generation at Walnut Log, Tennessee, near Reelfoot Lake. (Photo by Jim Johnson.)

Part 2

Confronting the Challenges

Chapter 9

The Northwest Territories

Trying to solve the problems at Reelfoot Lake became a high-profile project for the managers under my supervision in the northwest territories, but the projects on the Obion–Forked Deer often proved to be just as daunting and controversial for us from 1984 up through the present.

In fact, for TWRA director Harvey Bray, the demands of the Obion–Forked Deer projects began even earlier. It was in March 1974 that, as a result of the Akers lawsuit, the Corps of Engineers was authorized to purchase the thirty-two thousand acres of mitigation land in order to continue its work on the WTTP. The situation was neck deep in politics, and the state's waterfowl hunters anxiously awaited the outcome. With the agency obligated to a half-million license holders running out of public land on which to hunt and fish, Bray's vision was clear: to have these lands turned over to the state for development as soon as possible. It was precisely the kind of project for which Bray felt his talents were best suited. To do it, he wanted to have a crack lands-management biologist on the ground and obligated to the WTTP.

Frank Zerfoss: Supervisor of the Northwest Territories

Director Bray was prepared to search the country, if necessary, to get the person he wanted. So far, he had not found him inside his agency. Bray asked me to help recruit someone fitting his requirements. I had a prospect in mind—Frank Zerfoss. Frank and I were graduate school classmates at Southern Illinois University. He was savvy, ambitious, and had a winning personality; he could isolate problems and find answers. Moreover, he had a special interest in rivers and wetlands. After some negotiation, Frank left his assignment with the University of Kentucky Extension Service in 1974 and accepted the job. The territory under Frank's supervision involved seven of the northwest counties, the watershed for the Obion–Forked Deer.

Frank pursued the new job with characteristic enthusiasm. His expertise, however, in river ecology was no better than that of any of the rest of us. Very few in the wildlife profession knew much about wetland or river ecology during those years; instead, it was

Frank Zerfoss, the first supervisor of the "northwest territories," holds a brace of red-legged mallards taken from a naturally flooded bottomland hardwood forest. (Photo by Jim Johnson.)

something we learned together. Unfortunately, this need for a learning curve hampered Frank in one of his earliest assignments, which was to help design and then edit the future plan that Harvey Bray envisioned for the mitigation lands.

Once begun, the legal process involving the mitigation lands proved to be grindingly slow, and Frank was also assigned to supervise the wildlife management areas within Region 1's northwest district, including Reelfoot Lake, Gooch WMA on the Obion, and Tigrett WMA on the Forked Deer. During this early period none of us gave much thought about how the corps's West Tennessee Tributaries Project might affect the projects already in place. It was thought by most that Gooch's popularity with waterfowl hunters would blossom once the addition of mitigation lands allowed it to be more fully developed. This comfortable feeling would last only a short while, however. I was already beginning to ask pointed questions about the future of Gooch and the ecological carnage that it and dozens of other levee-dependent projects were having on the rivers. (And, as I shall discuss in more detail shortly, a seemingly hopeless situation at Tigrett WMA had developed that only reinforced my skepticism.) Shortly after he joined our agency, I met with Frank for a half-day briefing and mercilessly saturated him with information on the rivers, as well as my thoughts about the ecology of Reelfoot. The next day, we took a field trip to Tigrett to recapitulate what Speck Hurt and I had pondered over the last two years. Frank's confidence remained unshaken, and he searched for a starting point.

Gooch WMA, developed only ten years earlier, was beginning to show signs of levee deterioration and sediment build-up in the new ditches and on large tracts of hardwoods. Frank and I spent much time over the next two years trying to analyze these early warning signs. An eternal optimist, Frank saw possibilities and was determined to turn around the collapsing state of Gooch as well as Tigrett. But it would not be so easy. Zerfoss still had confidence in traditional thinking, and Carl Yelverton, the biologist who developed the plans for Gooch and Tigrett in the 1960s, left them in good order. Frank wanted to give Yelverton the benefit of a doubt and decided that we should try harder to keep his plans in place with a few modifications. Yelverton's plans were extremely well done, but they were guided by the old methods—levees, hot crops, and pumps. Like everyone else, Yelverton was unable to anticipate how levees and farm-oriented management for ducks would affect the rivers.

Tigrett WMA, as I have mentioned, had not been developed as Gooch had been—indeed it never would be—even though the acreage had been purchased at about the same time and Yelverton had drawn up plans for it. In large part, this was because Gooch had practically exhausted our money and human resources just to keep it intact. Managing wetlands had always been time-consuming and expensive, but this was especially the case when the lands were so squarely in the path of the river. Another delaying factor—and this also affected Gooch—involved the ownership of the land: flooding water knew no property boundaries, and the state did not own all of the land it needed for the WMAs.

This old waterfowl blind at Tigrett WMA is no longer useful since waterfowl no longer come to this artificial swamp. (Photo by Jim Johnson.)

Public versus Private Rights along River Corridors

The need for land around these wildlife areas brings up a critical subject that affected every tract of wetland the agency owned within the northwest district. Rivers are like public highways, and access is needed to use them. Those who are aware of the public's right to access navigable streams and public land often feel shortchanged. More frequently than not, private land is a barrier to these rights, hindering access to the rivers and often to public land.

In West Tennessee, the problem was worsened by the rising demand on the flood-plain. The farming of soybeans was a new movement here during the 1960s, and it did not slow down much before the 1990s. Consequently, more land was needed to meet the demand, and short-season crops like soybeans could be squeezed into the dry seasons when other cash crops could not, especially if levees protected the fields. Throughout this period the landowners were hell-bent on clearing fertile bottomland to have acreage for planting. During this same period, the state was just as determined to expand public lands for waterfowl hunting. Conflicts of interest without an enforceable floodplain plan were bound to occur. In this case, the requirements for TWRA's waterfowl management were not much different from those for normal farming practices: our interest involved ducks, and theirs involved agricultural profits. Neither one, unfortunately, was compat-ible with the natural functions of the floodplain. Thus, levees were developed helter-skelter throughout the length of the rivers, with no state-sponsored, comprehensive land-use plan in place. Nor were there prospects for one for the foreseeable future.

Gooch, Tigrett, and the proposed mitigation lands were at the center of this contro-versy, but the TWRA, like everyone else, was either oblivious to the problem or reluctant to tackle it. It was apparent that private rights and public rights were already on a collision course. If the conflicts were not resolved, future generations would not reflect kindly on our failure to take action; eventually, they would have to bear the costs.

The importance of the public's right to common property had been addressed before—in the case of highways, for example. But rivers, although they are navigable highways (and much more besides), have not been reckoned with in Tennessee in the same way as what we usually think of as highways. Navigability in the legal sense means that the public owns both the stream and the land beneath it. Navigable streams in the common sense mean that private ownership may include land beneath the streams but that the right to navigate them and use their water belongs to the public. This is the general rule that applies to the rivers in West Tennessee—that is, the public is required to negotiate purchase of the land beneath these streams in order to have all of the rights that go with the stream. In this respect, a legally navigable river is similar to a highway: private deeds often meet at the center of the road, or it may include the road, but the public has the usage rights and the right to develop and maintain it.

When public and private rights clash in the cases of rivers, a reasonable compromise usually involves just compensation for the easements or ownership of the river corridors, thus giving the public the access it needs and allowing the river to be properly managed with minimal conflict. Unfortunately—and this is especially true in West Tennessee—iso-lated easements will not solve the problem. The entire river corridor from its confluence to its upstream limits is required. Only then will equitable use of the landscape along rivers be rightly served.

Since this had not happen in West Tennessee, we found ourselves struggling with ownership problems at all of our waterfowl areas. This was the case at Gooch and Tigrett:

not enough land was available to accommodate the agency's plans without affecting private land (although the public's right to navigate the river was valid). Contention over a single tract of land can be enough to stall an entire project. Why landowners failed to complain more about *our* water on *their* land is still a mystery. Even so, many were willing to relinquish the land for a fair price, but timely funding to make such purchases was always a problem.

When Gary Myers replaced Harvey Bray in the early 1980s as director of the TWRA, he did not press for the purchase of all the land needed for Tigrett in the hope that the corps would eventually acquire the land as part of the WTTP mitigation project. After this, he hoped that the land could be developed. In hindsight, it was probably a blessing that all the land had *not* been purchased for development. If it had been developed, the managers would have been committed to perpetually sustaining something we now know does not work.

Useless Land Is Not the Best Wildlife Land

Wildlife agencies are always at the mercy of low budgets, and from the outset such financial restraints have, by and large, limited land purchases to cheap tracts that were worn out, too wet, or otherwise useless. But budgetary constraints aside, there is another reason that wildlife land has been limited to less-desirable acreage: the resistance of the farming industry. Farmers in general have opposed government purchase of prime or potential farmland, fearing that this will prevent its use for agriculture. And, naturally, the industries serving farmers—the implement, supply, and chemical companies—have also opposed trading land out of arable production, since it could mean fewer sales of their products. The Farm Bureau probably disdained government land purchases as well, out of fear that fewer insurance premiums might be issued.

Ironically, such concerns have been misplaced. Wildlife acreage generally requires cropland, and in the TWRA's waterfowl areas, cropland has, for the most part, been privately farmed through cooperative contracts. On those acres favorable to row crops, usually no more than 30 percent of the crops are left standing as food for wildlife. Taking prime farmland out of production is a rare event unless the land is too wet to farm; thus, there is little reason for anyone to complain about natural resources agencies purchasing farmland for wildlife. Nevertheless, overcoming old preconceptions is difficult, and making do with poor land has long been the lot of wildlife managers. Of course, wildlife cannot really thrive on such land; like crops, cattle, and timber, it usually survives best on fertile soil, if not well-drained terrain.

Thus, we have seen the failure of those management strategies I discussed earlier, such as planting "hot crops" like corn as food for waterfowl. Such crops, vulnerable to flooding, have failed more often than not and left us literally no room to do anything else on the land if we were to satisfy both hunters and farmers. To meet an area's objectives, every acre had to be utilized to provide reasonable carrying capacity for waterfowl. Consequently, most managers pushed the limits of their time and budgets to provide the crops. And as

I have noted, no matter how many seasons the crops failed, another attempt was planned for the following season.

By the early 1980s, there was little doubt that the TWRA had adopted the wrong policy for managing waterfowl in these river floodplains. And yet we seemed doomed to keep repeating such mistakes. Frank Zerfoss had begun to see the dilemma and the changes needed for a new strategy, but he reckoned with the old problem—tradition—even as the evidence against following the old methods grew stronger each year.

The Failing WMAs

Mistakes that had been made at Gooch WMA stood to be repeated at Tigrett. Willing to give tradition another chance, Zerfoss saw an opportunity when, during the early 1980s, the Basin Authority had to clean out the Forked Deer River canal that ran through Tigrett. The "clean-up" would result in a lot of excavated dirt piled up along the banks—a levee. Frank persuaded Richard Swaim, the head of the Basin Authority, to refurbish the levees along the main channel in a manner similar to what had been done at Gooch. The excavated dirt stacked from the main channel could be used to attach cross-levees and hold water cells to be managed for ducks.

Ironically, these levees were constructed immediately adjacent to old cells built by former duck clubs, and those had failed miserably, leaving dead oaks and stagnant swamps. But it was hard for us to accept that we could do no better than others had done. As had happened at Gooch, it was easy to be fooled by short-term success, which is a common occurrence in newly developed areas. Thus, at Tigrett, when fresh water was added to fresh ground, it produced an abundance of duck food, which resulted in amazingly productive waterfowl hunting for a few years. Then, as had happened at Gooch, the bills came due, and someone had to pay for the deteriorating levees, roads, and other facilities.

When our attitudes finally began to change, they changed in a permanent way. It had taken a century to cause the destruction we saw in the rivers. Further deterioration was difficult to imagine. When the damage at Tigrett occurred, there seemed to be no way to adequately describe or cope with it.

Facing Facts

Managing land for a single species had never been the right way to manage, but this too was a tradition. A better approach is to encourage and develop diverse wetlands, for such lands are more resilient and of better quality, even for ducks and duck hunters. And, of course, they provide a greater variety of wildlife for the use and enjoyment of other sportsmen and outdoor enthusiasts. Consequently, such lands gather more public support for wetland projects. We eventually found this to be true with such places as Black Bayou Refuge at Reelfoot, the management area where we first began to test new and innovative methods.

But through the first half of the 1980s, what we would learn at Black Bayou was still several years away, and Frank Zerfoss was obligated to deal with the plan for the mitigation land that Harvey Bray and the duck hunters so coveted. Many were caught up in the euphoria of having won thirty-two thousand acres of public land, but Frank and I had increasing doubts about the viability of the plans as they were developed. If they were carried out, we could be stuck for generations with a wrong approach. There was a lot more at stake than just ducks; the future of the entire river was in jeopardy. But we seemed to be alone in these concerns. The first committees devising the WTTP mitigation plan felt comfortable with it. Even as they grappled for a year or so with the final drafts, they failed to see the long-term consequences of what they were proposing. Meanwhile, everyone waited impatiently for the last tracts to be purchased so that development could begin.

The West Tennessee Tributaries Mitigation Lands Wildlife Management Plan was finally finished in 1983. If it had worried us while it was being developed, it concerned us even more after it was completed. With its proposals for levees that blocked the flow of water across the floodplain, it seemed like a surefire prescription for repeating the errors at Gooch. Frank remained optimistic, however, that somehow the plan's authors would recognize its shortcomings and revise it before development began.

Overall, however, Frank's enthusiasm was starting to wane. This began when Elmore Price, Frank's colleague in the eastern part of Region 1, announced plans to retire in 1983. A legendary lands manager who once supervised all of the TWRA lands in Region 1, Price had also been a staunch advocate of our efforts to change the plan. Frank decided to apply for Price's vacated position when the ongoing legal wrangling over the WTTP put the future of that project in question. Frank reasoned that if the WTTP were de-authorized, his current position could be reabsorbed and the original purpose for hiring him invalidated. Frank's hard-driving nature, however, did not fit the public relations profile expected of those who held the position. Frank lost his bid for Price's job.

The leadership of the region had changed in 1982 with the retirement of regional manager Wilbur Vaughan. Harold Hurst, the manager of Region 4 in East Tennessee, accepted a transfer to the vacant position and became the new manager of Region 1. Hurst found an entirely new set of problems than those in the mountains where he had worked before; now he dealt with waterfowl instead of black bears. Harold was a dedicated administrator and a fine person, but his arrival did not alter a disenchantment with my position as assistant regional manager that dated back nearly a decade. In 1974 the reorganization of the agency had given us an opportunity to revitalize the Region 1, but as I saw it, what little progress there was occurred in spite of the administration. Old mindsets prevailed that caused lethargy and an inability to implement the new goals. After a decade, even with the new administrative lineup, we gained very little ground. It was sluggish at best and, for me, a permanent source of frustration.

My roots went deep into the fertile turf of wetlands, and I had grown tired of being stuck behind a desk. I wanted to see dirt turned and physical projects implemented. This caused me to gravitate toward the field projects in Frank's district rather than tolerate any longer the drag of regional administration. Field biology and wetland management could produce tangible, measurable results that could be produced within months instead of decades or years, if at all.

By 1983 bureaucracy had finally taken its toll on Frank Zerfoss, and he decided to leave the agency. It was a sad but sometimes common fate for energetic, talented employees of government agencies. Frank's decision was an unfortunate and significant loss for TWRA. It was an even greater loss to me personally; I alone was left to deal with the rising uncertainty over the future of the rivers. Unable to change Frank's mind, I asked him to perform one more task before he left the TWRA—to accompany me in an appearance before the full staff in Nashville, where we would tell them what we had been thinking. They needed to hear that the agency was on a reckless path. The duty to tell them, like it or not, was ours alone. With Harold Hurst, I discussed our desire to address the TWRA staff, as well as my intention to apply for Frank's position. As Region 1 manager, he understood and agreed to support us, including my transfer if I were selected by the interview committee. After some consideration, we requested time on the TWRA director's next Nashville staff meeting to make our presentation. Hurst agreed to do this without any further discussion. Soon enough, I would find out that he had underestimated—or misunderstood—the intent and potential impact of our message.

The Director's Meeting

The meeting of the TWRA staff was scheduled for a Monday morning in February 1984, and Frank and I were the first item on the agenda. Director Myers and his staff had no clue that they were about to give judgment on one of the most critical issues of their careers: the future of TWRA's waterfowl program in West Tennessee. The tragedy, I suppose, was that no one, neither Director Myers nor any of his division chiefs, had been briefed about it. (And the timing, I admit, was not good: As I will discuss in chapters 12–14, projects at Reelfoot Lake had been a contentious issue during the previous year and were consuming considerable funds and manpower. Thus, the group in Nashville probably thought they had heard enough about West Tennessee already.)

The staff believed that things had gone fairly well with the WTTP litigation. The state seemed poised to receive the thirty-two thousand acres from the settlement, and this would fulfill promises made to a half-million anxious sportsmen. To tinker with it, to get it off course, would not sit well with the director and the chiefs. But we had no time for a soft sell; with Frank's departure upon us, time had run out.

Frank and I had dreaded this meeting for several months. It was going to be blunt and to the point. The staff had to know that, despite our best efforts, every waterfowl development project within the Obion and Forked Deer river floodplains was a failure. Not only had they failed to meet the agency's objectives but they had also begun to deteriorate from the moment they were built. And then there was the status of the rivers themselves: what would they say when we told them that we had contributed to the destruction of the waterways?

One thing the staff knew was that TWRA's waterfowl budgets seemed always to be exhausted. Nearly every year we required supplemental funding just to maintain annual operations. These facts alone should have been reason enough for them to listen. We hoped that they would see it from our perspective—that not only could more waterfowl wetlands be restored and saved but that the new objectives could also be accomplished through far better conservation ethics and at a lower cost. It might also provide a better share of the agency's annual budget to their projects. Healthy rivers, managed at a cheaper cost, could easily represent all that the agency stood for. In effect, such a vision could become a way of selling the value of the agency's good work.

"Yes," the chiefs might say, "but what about the powerful special interests TWRA would need to face?" We would agree that this could be a problem. However, vested interests had probably driven most of our lives as state administrators, and it was time to challenge this—for us and all Tennesseans. The chiefs might also mention that confronting special interests could endanger careers. We were not naïve. We had seen the most honest efforts reduced to silence by vested interests. But we believed that all the public needed to do was reject vested interest. That, plus patience and more education, could very well secure the public's backing for these new ideas.

Once we gained the chief's support for our new vision, we hoped to follow up with talk about specific projects. First, we would mention Gooch and Tigrett, and the failures, expense, and best solutions for dealing with these flood-prone wildlife areas. These operations needed to be relocated just beyond the annual floodplain—not *in* the floodplain. Perhaps the current operations could be moved beyond the five- or ten-year floodplain, enough to satisfy the need to raise "hot crops" for waterfowl. Funding and willing sellers would be issues as always, but our strategy was not that farfetched. Funding and willing sellers had been found before.

Second, we would mention that public control of the river corridor could provide tremendous wildlife opportunities and, at the same time, solve most of the conflicts that had brought us to this juncture. For one thing, the demand for channelization could be minimized by public ownership. Moreover, the operations expense could be greatly reduced, as management became more efficient and effective. Natural drains could be restored, and the floodplain would begin to function favorably again, and this was something the farming community could not help but favor. In the end, the change would benefit everyone. But the most significant gain—the most long-term bonus—would be the enhanced function of the rivers. An enormous burden would be lifted from the shoulders of the public, and farmland would no longer suffer the carnage of flooding. Far more prime habitat would be gained for all species of wildlife and fish. Consequently, wildlife users would have many more opportunities. The potential benefits were staggering.

The thirty-two thousand acres of mitigation land could be managed similarly. There was no good reason why this view should not be accepted by the staff, nor was there any reason why the public would not accept it. Once they saw that the alternative was untenable—that the agency's wildlife areas would eventually be destroyed along with the future potential for waterfowl in this section of the state—there was no other good choice. This was to be our message.

The somber chiefs settled pensively around the large, U-shaped conference table. I hoped they would at least listen with open minds. Perhaps a meaningful dialogue would develop. Maybe they could see that TWRA's oversight in management was flawed not because the objectives were wrong but because our methods and policies had been misguided. I launched into the presentation. It was over before it had hardly begun. In less than five minutes, Myers and his staff had heard enough. To them, we were way off the mark. Our ideas, in their view, threatened the one bright light for the future of waterfowl hunting in West Tennessee: the potential for the state to gain thirty-two thousand acres of mitigation land.

"Why are you telling us all this now, Johnson?" Myers interrupted, his jaw set and his face red with anger. "Where were you guys in the planning process?"

The worst had happened. There would be no way to rise above this response, and to a great extent, I understood Myers's reaction. A long process, a large expense, and a massive effort had gone into the J. Clark Akers lawsuit. Large obstacles had been overcome to gain the concessions issued by the courts for the mitigation land, and the state was about to receive a huge acreage. A lot was at stake. But, to Frank and me, it could be trashed by a fatally flawed mindset for managing these lands. The irony was that no one wanted success more than we did. Certainly we wanted to see public ownership of the river floodplains. Logically, it was the best assurance we had to protect at least part of the river. And there was the public need: certainly the citizens needed these lands for recreation, even though they were needed more to help save their river.

However understandable the response we received may have been, the TWRA chiefs should have heard us out. Had anyone considered that the concession for the mitigation lands would require the state forever to contend with the devastation already caused by channelization? Did they understand that we had accepted a land package that would not work, that it would perpetually drain the agency of its manpower and funds? It apparently had not crossed anyone's mind.

Unfortunately, our few remarks caught our direct supervisor, Harold Hurst, off guard. "You have blindsided me," he said, stunned. "I didn't know you were going to say this." It was as if we had never mentioned the problems. As TWRA administrators saw it, we had come to condemn the projects and perhaps them as well. I wanted to explain that I was as responsible as anyone for the problems mentioned; they were projects under my immediate supervision. But to propose such an enormous reversal in vision and strategy was too much, too fast. Any old "River Rat" might have been far more convincing than we were—but it was too late. The meeting ended without further discussion or comment, but there were a lot of bruised feelings, among them our own.

Frank and I departed for West Tennessee, bewildered, dejected, and disgusted with the outcome. Not until the next year would the subject be mentioned again. A year, however, changed some minds. Myers and Hurst had apparently thought more about the meeting. Before the year was out, they came to visit our projects. Zerfoss had gone, and I had replaced him. Ralph Gray, the area manager, and I accompanied Myers and Hurst to the swamps. They asked hard questions. Before they left, they heard most of what Frank and I had intended to say at the staff meeting, and it was clear that their attitudes were different. Afterwards, Myers and Hurst supported not only the Reelfoot projects but also the restoration of thirteen miles of river at Tigrett WMA. The question was, now that they had begun to see the picture, how long could they stand by projects with such a controversial profile?

At one point, Ron Fox, the TWRA field director, encouraged us. His ideas and ours had diverged sharply. Nevertheless, he and I met in August 1990 at the Holiday Inn in Jackson to discuss slight changes in the WTTP mitigation plan. He advised me that the corps had announced that it was prepared to release the first thirteen thousand acres of the land to the state. He wanted to know whether we were ready to redesign Gooch WMA. This was an encouraging question, since he had been a staunch supporter of the corps's development plan for Gooch WMA and had opposed our river restoration proposals.

The next question was more sobering. Fox wanted to know whether we needed to exclude the current mitigation development design, since it would affect flooding on private lands. I was unsure how to take this change of attitude. We had made this recommendation at the director's staff meeting in Nashville, but it was rejected.

To this day, however, I have no idea why we had that conversation in Jackson. By the following summer, the field director's attitude was different. Fox downplayed the Tigrett floodplain restoration project at a biologists' meeting at Paris Landing State Park. He indicated without explanation that we no longer needed to consider that plan. The corps, it seemed, would be taking charge of river restoration. His attitude, and what he knew that no one else did, remained a mystery for a long time.

After mentioning Fox's ambivalence to Myers sometime later, he offered his condolences. "Eventually they will come around," he said. He went on to outline other things on his mind. As soon as we finished the Reelfoot project, restored Tigrett, and moved out of Gooch, he said, he had hopes for the TWRA to acquire thousands of acres along these rivers—from the headwaters to the Mississippi River. We looked at the maps and contemplated the thin blue lines. "We'll do this like you guys have recommended," he said. He went on to say that "we'd do it right." Myers was always optimistic despite the troubles; there was conviction in his face, the kind that comes when administrators are determined to stay the course and do the right thing. It was something to build on.

Chapter 10

The Tigrett Floodplain Restoration Plan

In 1980 the Tigrett Wildlife Management Area had been waiting for help going on a good three decades. It was not easy to forget that, during the 1950s and on into the early 1960s, good stands of bottomland hardwoods still could be found in the lower units around Stokes Creek and on certain tracts across the river on the north side. Fair hunting for small game and ducks was still possible during those years. Fishing on Tigrett from the Ro Ellen boat launching area on the south side was very popular in the swamps of the old river bed. But those days were gone; in fact, they were practically gone by the end of the 1960s. Any successful duck hunting in the area was found eight to ten miles farther upstream around Mills Slough, where a few trees still stood and the margins of the floodplain could dry out enough for native food crops to grow.

Stokes Creek farmers complained about the inordinate flooding on their soybean fields even as they failed to address the root of the problem—excessive sediments caused by soil erosion and the loss of a river. They, like everyone else, were addicted to the subdue-and-control mentality that had created this mess in the first place. It had reached crisis proportions by the mid-1980s, however, and the farmers wanted relief, which they saw no way of getting without government help. But those of us they called on were stuck with the excavated ditch—the anti-river. The labyrinth of levees and artificial channels crisscrossing the floodplain had brought drainage to a halt and left nothing but stagnant swamps and dead trees.

The farmers' complaints turned out to be the catalyst we needed to realize our hopes of formulating a floodplain restoration plan. If such a plan could be developed and implemented at Tigrett, it could serve as an example of how the entire Forked Deer River should be managed. We had no doubt that the plan would go a long way toward solving most of the river management issues we now faced. The intent of the Tigrett plan was to free up what was left of the natural drains and perhaps get the stranded parts of the old river functioning again. Without full restoration of the river, continued maintenance would be required, but partial restoration was better than nothing. It would at least give us a chance to show that natural hydrology was the best answer. We planned to leave the "Big Ditch" in place and tackle it at another time, since at this stage the mere mention of filling it in would almost certainly cause many to doubt our sanity. With these thoughts in mind, I set to work with Ralph Gray, the Tigrett manager.

Assessing the Problems

In developing the Tigrett plan, the first job was to identify the principles as clearly and concisely as possible—to keep it simple, in other words. We wanted to restore the Tigrett floodplains as fully as was practically possible, guided by the evidence of natural hydrology. It might include former drainage, or it might not, since the topography of the floodplain had changed. Next would come the inventory. We had to identify the physical features: the routes of all drainages, the levees, the topography, and dozens of other details of logistics and construction.

Our "reference river"—our point of comparison—was the Hatchie, which formed the next watershed south of the Forked Deer. The contrast between the Hatchie and the Forked Deer was remarkable. The Hatchie, relatively untouched by channelization, was in far superior condition; it was the only natural river along the Mississippi for more than a hundred miles north or south. Our guess was that the Big Black River upstream from Vicksburg was the nearest stream whose health could compare to that of the Hatchie. "The Hatchie is the best river system in West Tennessee. I've been on most of them," Joe Guinn, a nature photographer, told me in an interview. "And the Hatchie is the best system in the Lower Mississippi Valley, from Cairo to New Orleans."

In the summer of 1984, I sketched the first rough draft of what would eventually become the "Tigrett Wildlife Management Area Floodplain Restoration Plan."[1] The draft represented practically everything we had learned over the years about Reelfoot, the Hatchie, and the other rivers. When the draft was finished, neither Ralph nor I would have bet a two-dollar pickup truck that the plan would go any further than our Region 1 office, let alone set an example of how to manage a river. Nevertheless, we felt we had to put our ideas on paper.

The TWRA Nashville staff had already shown that it was not prepared for a project with such uncertainties. Like the duck hunters and the landowners, they found it easier and less complicated to make do with what we had and live with whatever things seemed to work. A Tigrett advisory group, formed in 1990 and known as the Tigrett Technical Committee, was also skeptical. This group, which I had started to help us work through the many challenges of the proposed Tigrett restoration, included representatives from various wildlife, environmental, and conservation agencies and organizations, both private and public. It was not reassuring to find that these noted wetland experts were skeptical about what we were trying to do. They wanted us to be successful, but we were treading in uncharted territory. If they had known our long-range intentions—not just to effect the restoration of the Forked Deer and the Obion but that of all West Tennessee rivers—they might have been even more skeptical. It was best not to mention it. The full scope of our intentions would not be known until 1991, when we would release a video explaining the subject to the public.

As a group, the Tigrett Technical Committee was not a risk-taking body. The prevailing opinion was to leave the river alone. So many modifications had been done that it seemed impractical to expect that we could restore it to anything resembling a natural state. Some wanted restoration to start in the upper watersheds, or at the mouth of the river, but not in the middle. The worst objection, unfortunately, was a majority opinion: most committee members believed that the artificial swamps should not be touched. I could understand this position since, at first, I had thought the same thing. But now I saw it as a real problem. It was clear to me and my staff that such unnatural wetlands contradicted—in a monumental way—all that we knew about the essential principles of a natural river. How could one expect the floodplain to drain with such obstructions directly in the path of the river? I knew that any structures in the floodplain had to be compatible with the needs of the river.

The Tigrett Technical Committee made some good points. It indeed would have been better to start restoration somewhere other than the midpoint of the river. But that was where Tigrett WMA was located. The state did not own the land at either end of the river, let alone to the edges of its floodplain. Some four thousand acres and a thirteen-mile segment of river in the middle were all we had at this juncture. Could the floodplain be restored in the areas the state owned? It could probably not be done well without cooperation from local landowners, but perhaps we could do enough—such as restoring and reconnecting the old meanders of the river and removing levees and other major blockages within the WMA—to demonstrate what was possible.

From the air, Tigrett looked like a glistening marsh, an apparent mecca for aquatic life and waterfowl. It only fooled our critics. Black willows, buttonbush, and giant southern smartweed dominated these stagnant swamps—favorable for brooding wood ducks but terribly deficient for mallards and other dabbling ducks. The seeds of grasses and other annual vegetation preferred by dabbling ducks were found only at the edges of the swamp, where the land could dry out before reflooding. The oaks were long gone, and even the buttonbush and willows were dying back, leaving clear, open waterholes that were largely useless for fishing or hunting. As a result of the sedimentation and flooding problems brought on by channelization, very little water flowed through Tigrett at normal summer river stages. In a 1990 study comparing artificial swamps to well-drained bottomland hardwood forests, Neil Miller of the University of Memphis noted, "Oxygen level recorded in the perennially flooded marsh area [the artificial swamp] would preclude this area from serving as a viable game fish habitat." The only fish found in the artificial swamps, according to Miller, were a few mosquitofish, *Gambusia affinis*. This is a nongame species, a small surface-dwelling fish that attains a maximum length of only five to six centimeters.[2]

Miller's work helped legitimize our own observations about the character of artificial swamps. His work on the absence of oxygen in these areas helped confirm our basic point—that restoring free-flowing water was a key to recovering the swamps and reestablishing the species of fish once common in these river systems. But this information did not satisfy our critics, who thought that the problem had to be more complex than basic biology or hydrology.

Of necessity, our plan for Tigrett dealt with numerous construction problems, one of the worst of which was an abandoned Gulf-Mobile railroad track that crossed the middle of the WMA. The structure barricaded almost the entire floodplain. Only five narrow gaps in the railroad embankment on the south side of the channel allowed floodwater to pass through the earthen embankment of the railroad bed; this was less than a tenth of the openings actually needed when the river was at flood stage. Breaching the roadbed could increase positive hydrology.

Highway 412 at the downstream border of Tigrett also functioned, in effect, as a levee. Open-span bridges limited the flow to bridge openings at the main river channel and at a location south of it where the natural river once flowed; neither of these outlets could accommodate the volume of water flowing down the river. Consequently, the entire thirteen miles of the Tigrett floodplain upstream were adversely affected by rising water.

The design justification for the earthen portions of the roadbeds was simple economics: it was cheaper not to build bridges. Dirt roadbeds cost about half of what a similar length of bridge construction cost. However, the damage these roadbeds have caused with upstream flooding was never assessed—and this despite the fact that highway departments (as well as railroad companies) can be liable for damages when their structures wreak havoc to private property. But neither the Tennessee Department of Transportation nor the state's political leaders showed any interest in solving this problem, since doing so could burden the state's budget. It was "ostrich economics"—sticking one's head in the sand to avoid a true, long-range analysis. Probably the surest death sentence for a river restoration project is the mention of a conflict with a public highway. Perhaps the best way to overcome this obstacle is to complete an honest cost-benefit analysis, one that includes direct and collateral damages the highway causes to the ecology of the river and what the long-term savings might be if the roadway were removed or modified. However, in drafting the Tigrett plan, we could do no more than make slight mention of such needs and difficulties; otherwise, it would surely have doomed the entire proposal.

The levees formed by the excavated ditch that replaced much of the original river corridor through Tigrett, as well as the levees built by the old duck clubs, constituted one of the most pressing issues we had to address in our plan. Excavated material was stacked eight to ten feet high alongside much of the straight channel, which had first been constructed around 1916. The smaller duck club levees met this channel embankment at right angles in several places. Where these ran across the floodplain to higher ground, or to another levee, they created compartments like those on Gooch WMA. When the outlets to these compartments were clogged, usually by beavers, they trapped water and blocked the flow that, in natural systems, went down the floodplain. Each enclosure created an artificial swamp, not unlike those that were created when the old river meanders were cut off by channelization. These levees also had to be breached.

Sediments were a special concern. Very little could be done to remove them from the floodplain (not that they could not be removed from the channel itself), but slowing them from further entering the river was essential to a successful restoration project. Considerable progress was being made toward this end by government-sponsored soil conservation programs, such as those promoting no-till farming practices. Some had insisted that soil erosion had to be stopped before any restoration work could begin, but this was an unacceptable excuse, in my opinion. Continuous progress toward the reduction of sediments entering the river was enough to keep moving forward.

The accumulation of sediments, if one judged by the coverage of tree roots, was as much as seven to ten feet deep at the upper and lower ends of the old river, and this further compartmentalizing of the floodplain created even more artificial swamps. These swamps became as still as farm ponds after the river fell to its normal stages. Stagnant swamps held perennial water over all but a tenth of the floodplain. Restoring the natural hydrology of such swamps could show positive results as early as the next growing season: trees would sprout, vegetative growth would return, and wildlife would return to the river valley at the same rate. Technical experts tended to act with excessive caution and rarely agreed with this simple approach. But those of us seeking to restore Tigrett did not agree. We felt it best to gather all good opinions together as quickly as possible, and to move all measures forward concurrently. Of course, we never ruled out reasonable and sensible caution.

The initiative and aggressive thinking revealed in the Tigrett recovery plan gave us a reputation as "loose cannons"—wild-eyed managers unleashed on something we knew little about. But we had done our homework better than our critics knew. We were well aware that some of our proposals were untested (after all, this was a demonstration project), but the best model to follow was in our backyard: the Hatchie River. We believed that following the lessons from that relatively natural stream put the odds of our being right greatly in our favor. From our perspective, anything proposed or done to improve the natural flow of the river was the right action. One thing we could not accept was the status quo—making the same mistakes over and over again without gaining or learning much. One other thing was sure: what we proposed was easier and less expensive, and promised more gain, than anything else that had ever been done. Even limited success could prove our point, and this was reason enough to continue with our plan.

The Stokes Creek Problem

The situation at Stokes Creek was the key factor that helped keep the Tigrett floodplain restoration plan alive from 1985 through the early 1990s. Dealing with the farmers' problems at this creek was a major challenge because the outlet went into one of the largest swamps on the south side of the wildlife area. The Stokes Creek watershed lay near the community also known as Tigrett. The creek was much like hundreds of others along the river, which is to say that it performed poorly.

The trouble began years earlier. Stokes Creek had once entered the old river, but the section of the river at that juncture was rendered useless by the artificial river channel. Stokes Creek, therefore, had no outlet; it ended in the quiet, stagnant waters of the swamp that covered the old river meander. Sediments from eroded soils of the watershed accumulated throughout the lower reaches of the channel as the velocity of the stream decreased. Beavers immediately took advantage of the sluggish flow across the floodplain and built too many dams. The Basin Authority's solution during the 1970s was to bypass this problem by digging a lateral ditch from Stokes Creek, around the swamp, and upstream to the new human-built channel. Ditches like this one made so much sense to those wanting quick solutions that it was virtually impossible for them to think any other way. It was quick, simple, and to the point. But it was successful for only a short time—a week, a season, a year, or perhaps five years. It never lasted as long as the farmers or the builders thought it should, so the response was to repeat the process.

When R. T. Fussell moved to the Stokes community in southeastern Dyer County in 1946, the creek never flooded. "When I built this house, I had no idea this was going to happen," he said in a 1996 newspaper interview. "We had better drainage then."[3] He planted a few row crops and was said to have "raised the prettiest herd of short-horned cattle in the region." But some years earlier he was forced to sell his cattle, and after that he exclusively raised row crops. He did well by producing as much as two bales of cotton per acre.

"Today," Fussell said, "much of that land raises little more than willows. . . . Beans are about the only thing I can gamble on these days. . . . You're lucky to get a crop in. Ninety percent of the time, you lose it. . . . One crop out of four, you might plant it and gather it."[4] Across this region, the income from this land was barely enough for farmers to pay their property taxes. Some could not. The failures at Stokes Creek and Tigrett WMA were classic examples of the consequences of channelization that can still be found across West Tennessee.

If the condition of Stokes Creek did anything, it encouraged Director Gary Myers to support the renovation of the floodplains on that side of Tigrett WMA. The landowners assessed correctly that drainage on their acreage had stopped because the WMA would not drain—the result of mismanagement. Now it had become the landowners' problem as well as ours. Director Myers could see their point, and the landowners quickly gained an ally.

Representative Harold Holt of Dyer County became involved. Area manager Ralph Gray, regional manager Harold Hurst, and I met with him on May 2, 1990, to hear his version of the farmers' complaints. The legislator was an elderly and congenial gentleman. After our discussion, he realized that the TWRA wanted the same improvements at Stokes Creek as the landowners, and this rare agreement was a politician's best news. Afterwards, Representative Holt pledged his support for the Tigrett plan and called Myers to express this view.

On December 9, 1990, Myers, Dan Sherry, the statewide environmental biologist, and Dick Stark, the agency pilot, scheduled a flight over the area to firm up the Tigrett plan. Within thirty minutes, Myers was satisfied that he had seen enough. After we landed, he was the first to speak. "Can we fix it?" he asked.

"Sure, if we have the stamina," I responded.

Ralph Gray and I had stewed over restoring Tigrett for more than a year now, and Myers's question was the one we wanted to hear. The director was encouraged by our response, but he did not consider our draft plan sufficient.

"Then, let's do it," Myers said. "I want you to lay out a design for Tigrett like you did for Reelfoot." Then, he added, "Don't do it just to resolve the Stokes Creek problem; do it for all of Tigrett."

Myers enlisted Dan Sherry to assist me with the tall order of preparing a complete plan. Dan, he said, was available to do whatever was needed and that I should let him know about any additional help I might need, or land to implement the plan. It was an uplifting moment. How could such a positive outlook not energize anyone? Whether it would work or not was a big question, but we would find out.

Dan Sherry, the TWRA environmental biologist who helped revise the Tigrett WMA Floodplain Restoration Plan and helped guide the required permits through the environmental review process. (Photo courtesy of TWRA.)

Later, Director Myers went on public record in support of our work at Tigrett: "Our political leaders recognize that there's a real-world problem as it relates to the flooding in these river bottoms. In our view, these problems should be addressed in as scientific and rational way as we can. Eventually, you need to stop talking about it . . . or should we do nothing? In my view, we have a management area in need of management, and we ought to do something about it."[5]

Our goal was to restore all of the natural hydrology in the southwestern section of the Tigrett floodplain—about ten miles of it. If it could be done—and this was one thing I was absolutely certain of—the floodplain would rapidly recover, and farmers at Stokes Creek would realize immediate relief. Only then would the plan be taken seriously. Then we would be one step closer to the ultimate goals, changing the whole paradigm for managing West Tennessee's rivers. So, Stokes Creek became, in a way, the impetus for restoring Tigrett and keeping our long-term goals alive.

The first step in implementing the plan was to gain political and public support (a lesson, as we shall see, that was learned painfully at Reelfoot Lake). Had we been able to develop broad public support, we might have counteracted special interests. Starting in February 1991, several public meetings were held at Walton Guinn's cotton gin. Howard Hurst, Ron Fox, Ralph Gray, and I represented the TWRA. With only a rough set of notes and charts, we outlined our solutions for the problems at Stokes Creek and the Tigrett WMA. Representative Holt attended one meeting along with local farmers Jimmy Barbour, Waldon Guinn, M. V. Williams, and a dozen others in the Tigrett community. For the most part, these farmers had not overcrowded the floodplain, although they had farmed to the very edge of it. Some of their fields had been successfully farmed for genera-tions. But things had changed. Channelization had set them up for false expectations and even greater drainage problems. Now the flooding worsened each year.

Laying out the basics of the plan, we made our case that the best thing for Stokes Creek and the farmers was to restore the natural hydrology and habitat of the floodplain in Tigrett WMA. To do this, several techniques were proposed: breaching internal levees, reincorpo-rating old river meanders, reestablishing incoming tributaries into discrete channels, con-structing low-level terraces above the annual flood-risk zone to hold water for waterfowl, reforesting former bottomland hardwood habitat, and reducing incoming sediments.

Floodplain restoration was a strange concept to these men.[6] Without thinking about how well nature could provide for their needs, they must have looked upon our plan as some sort of "tree hugger" project. But some had already begun to suspect that channel-ization was not a cure-all. Guinn was one. He had seen the mess it had caused. He also knew how nature managed the old river, and he agreed that the natural river was the best solution, if only it be reestablished.

"These bottoms used to be timber all in it," Guinn would recall. "You used to go down [in the river bottoms] and hunt squirrels and other game in the fall. The ducks would come and stay. . . . They would feed on the ridge where the acorn trees were. Then they'd go to the sloughs, and back to the ridges. They would not leave. Now, the only things they have are the fields to get any food. They light in the fields, stay a day or two, and leave. It's because of all this poor drainage. . . . It's killing all of this timber. We are like seven inches below rainfall this year, and we still have water all over these fields."[7]

The Stokes Creek farmers listened patiently, though some scratched their heads in thought. Although they had doubts, most supported the Tigrett plan from that day for-ward. Several meetings followed this gathering, some at the cotton gin, others at the courthouse and other meeting places in downtown Dyersburg. None of the meetings were adversarial, and many other Dyer County residents eventually followed the Stokes Creek farmers in support of the Tigrett plan.

Outside of funding and the necessary support, two constraints began to surface. One involved questions of what to do with the Big Ditch. The other centered on how best to cope with unabated sediments transported from upstream. These were the testiest issues.

Luckily, we had a new model to look at: Florida's multi-million dollar restoration plan for the Kissimmee River.[8] Florida's work left no doubt that the artificial channel had to be filled in; otherwise, restoration would not work. For the overall watershed, excessive sedimentation could be reduced to a tolerable level.

Political support also began to grow. A few weeks after the Stokes Creek meeting, Dan Sherry, our statewide environmental biologist, and I met with Director Myers at his office in Nashville. Myers had talked with the governor's office about our restoration project. Governor Ned Ray McWherter had a special interest in the WTTP, and he also had sympathy with efforts to restore the rivers. Mainly, however, the governor thought that if the plan worked, it might stop some of the squabbling over the rivers and ease the complaints coming to his office. McWherter had clarified his feelings about the WTTP and channelization two years earlier. "My position with respect to the West Tennessee Tributaries Project is also very simple," the governor wrote in a February 29, 1988, letter to the Wildlife Management Institute. "If the necessary flood relief can be accomplished in an environmentally sound way, the project should go forward. Additionally, and importantly, the side benefits of conservation, improving wildlife habitat, protection of hardwood forests and sound agriculture practices should be a key element with respect to the project's future. If these items cannot be accomplished, the project should not go forward."

His message was straightforward. By 1990 Governor McWherter was showing a special interest in the flooding problems at Stokes Creek. The word was that he thought we had made a step in the right direction. It soothed Myers's concerns as well. By November the plan was in the second draft and available for review by the Nashville staff.

Chapter 11

Rejection of the Tigrett Plan

To my great disappointment, the Nashville TWRA staff disagreed not only with the details of the draft Tigrett Floodplain Restoration Plan but also with its underlying concept. And without that group's support, we would make no progress. They made no recommendations but only said that they could not give the plan their blessing. Perhaps the plan had become too steeped in politics and had fractured too many egos for acceptance. Some administrators still chafed at what they perceived as threats to the mitigation land from the Tigrett plan, particularly its call to do away with levees and artificial swamps. The staff simply could not fathom that the swamps should be drained. After all, they said, these developments were exactly what was planned for the mitigation land and very similar to the structures in our existing waterfowl areas. This was true enough, but in those cases we tried not to leave the water standing in them all summer. Of course, these WMAs often failed and were part of why we saw the need to change our management policies.

"No net loss of wetlands" had become the thoughtlessly overused—and misused—slogan; it did not seem to matter whether the wetlands were good or bad. Aubrey David McKinney, the state's chief of environmental services, reasoned that it was better to have something than nothing. The artificial swamps at least had a few fish, he said. His argument had been used by others, but it did not add up. At best, its tenets served only a few wildlife species; at worst, it was a shame because the practices it advocated had contributed significantly to the destruction of the rivers.

In general, the TWRA staff stood its ground and preferred we leave the river alone. But Director Myers and regional manager Hurst held fast to their earlier conclusions about Tigrett. "The way I see it," Myers said, "the management area needs help. We can't just sit around and do nothing." His argument, in my view, had more merit since he had seen the rivers from one end to the other, and the others had not. The river's deterioration had been addressed to his satisfaction in the Tigrett plan. Nevertheless, he asked us to go back to the drawing board and reconsider the plan in light of the staff's comments and those that had arisen during the environmental permits process. Myers wanted the revised plan by December 1. He hoped for a compromise, but I feared that foundational principles might be lost.

Biologist Dan Sherry and I went into seclusion at Dan's condo in Destin, Florida. Without the distraction of telephones, we worked on the plan for a week. A rough copy was delivered to the Nashville office on November 28, 1990. Ken Arney, the chief of forestry, and Steve Patrick, the assistant chief of wildlife, helped organize the final draft for typing. We reviewed the work the following day with members of the Tigrett Technical Committee. At 4:30 p.m. on December 1, this version of the Tigrett Wildlife Management Floodplain Restoration Plan was placed on Director Myers's desk.

New objections arose from the TWRA staff and some on the Tigrett Technical Committee. The most frequent criticism this time was that the plan did not meet the technical standards for good civil engineering. Technical standards? They had missed the entire point. This plan was not based on principles of good engineering but on good natural resource science. "Good engineering" without regard for the needs of the rivers had produced the tragedy we were now facing. It was time, we felt, to get the priorities straight, with the leadership coming from the wetland managers. We knew that engineering had taken men to the moon and back, built bridges across the Mississippi, and erected skyscrapers in large cities. But engineers were not wetland managers. There would be plenty left for them to do once we had gotten the natural resource science right. But the review committee had thrown out the proverbial baby with the bathwater. Myers asked for no changes himself. He liked the plan but wanted time to think it through.

We still had to address the Big Ditch, which was an engineering issue because engineering had created it. No one living had seen the original meandering river before chan-

Members of the Tigrett Technical Team—including Dan Eager (left) of the state's Water Pollution Control Division, Dan Sherry of TWRA (center), Ben Smith of the Governor's Office (second from right), and the author (right)— discuss hydrology problems at Stokes Creek. (Photo by Kathy Krone.)

Rejection of the Tigrett Plan

TWRA agents Jeff Joyner, Harold Hurst, Jack Colwick, and Jerry Strom, along with USFWS agent Don Orr, inspect the artificial swamp at White Oak WMA. (Photo by Jim Johnson.)

nelization changed it. Thus, it seemed inconceivable that the river could function without the design the engineers had created. If the staff had reviewed the South Florida Kissimmee River restoration project, there would have been no need to convince them, but no one bothered to consider it. In Florida, engineers addressed the problem of channelization by filling in the counterfeit river and allowing the principles of native rivers to reestablish themselves. But our critics, like the farmers, assumed that the main ditch would always remain as a backup, a "security blanket" just in case nature failed.

So the review by the Nashville staff went about as expected—not very well. Other agencies and the Tigrett Technical Committee required additional changes to the plan. Myers requested an environmental assessment to address the various concerns. This grueling task would once more fall to Dan Sherry and me. It was a job we loathed but a necessary one.

Meanwhile, we hoped to bolster our case by finding a local demonstration project to confirm the conclusions in the Kissimmee plan. The opportunity presented itself at White Oak Creek, a minor stream about sixty miles southeast of Tigrett in White Oak WMA and some eight to ten miles southeast of the town of Henderson. Even as controversy loomed over the Tigrett plan, a 404 permit application was granted without much notice, and it outlined a plan to restore a section of White Oak Creek on the WMA using principles similar to those contained in the Tigrett plan.

Within a short time, the White Oak permit was issued. During its short life, the restoration project was remarkably successful, showing what could be duplicated on other streams affected by channelization. It lasted only a few months, however. About the time we were prepared to publicize the success of the project, our example vanished. A local landowner adjacent to White Oak WMA wanted more flood relief than he was due. After threats of a lawsuit, the man pressured Region 1 manager Hurst into removing some blockages in the artificial channel for temporary flood relief. Then, to no one's surprise, in one major flood event after the blockages were removed, the success of the restoration project was obliterated. The natural channels filled with sediments, and the floodplain immediately reverted to the former swamp conditions. If nothing else, this episode supported the Kissimmee River conclusions. But it was not enough to change the minds of our critics. Some on the review committee thought that the Tigrett plan was apt to cause more environmental calamity than it was worth. In general, their advice was to abort the plan. Consequently, more changes were made, and the size and extent of the project was reduced to satisfy them.

The Technical Committee and the Artificial Swamps

The purpose of the Tigrett Technical Committee was to guide the Tigrett Floodplain Restoration Plan. In addition to the Stokes Creek problems, the controversy involving artificial swamps strongly influenced the formulation of the Tigrett Technical Committee, if not the plan itself. An important step in the development of major natural resource projects is to involve a committee of people with expertise and interest in the project. These people are usually experts in their fields or key people in the local community. I had chaired one of these committees at Reelfoot Lake to address the drawdown there (see chapters 12 and 13), and it was fairly successful. Success requires a committee of a few key leaders and experts with genuine interest in the overall project. It is the only way to help keep focus on important issues and avoid the traps of miscommunication and personal agendas that interfere with general goals.

The first formal session of the Tigrett Technical Committee was held on July 30, 1990, about six months after the first draft of the Tigrett Floodplain Restoration Plan was completed. Participants were environmentalists or environmental regulators who frequently commented on environmental permits. The first members were Daryl Durham and Dan Eagar from the Tennessee Department of Environment and Conservation (TDEC), civil engineer Don Porter from the Tennessee Valley Authority (TVA), biologist Bob Ford from the Tennessee Conservation League (TCL), and environmental biologists Dan Sherry and Jerry Strom from our agency. Eventually, two of them became particularly savvy about the requirements and needs for river wetlands—Dan Eagar and Jerry Strom. Even today, these two conservationists actively guide the conscientious management of West Tennessee rivers. Dan Sherry, a member of the TWRA director's staff in Nashville, remained a key figure for the next fifteen years. His specialty was environmental regulations, and hardly any wetland projects passed through our agency without his close scrutiny as to their compliance with regulations.

This is not to say that others did not also contribute in significant ways. Bob Ford, for instance, eventually left the TCL and became more involved with his specialty—neotropical songbirds—with the U.S. Fish and Wildlife Service at their Patuxent Wildlife Research Center in Maryland. In this position, Ford continued to contribute to long-range wildlife resource goals in Tennessee. More important, Ford returned to West Tennessee in 2005 to become the service's regional waterfowl biologist, a position recently vacated by the retirement of Don Orr.

All of these participants had special talents that could and did contribute to the prudent management and use of river wetlands. Many had been involved since the earliest stages of the Tigrett project. But this is not to say that these particular members agreed with my concepts of wetlands. In fact, we often disagreed—rarely was a 100 percent consensus reached. The main reason was that we were involved in a young science in which none could claim to be the ultimate expert.

Later, individuals with entirely different perspectives joined the group: James Byford and Wintfred Smith, professors at the University of Tennessee at Martin; USFWS refuge manager Randy Cook; and Sam Anderson, a member of the UTM staff and a TCL director. The members of the Tigrett Technical Committee were highly respected in their fields. The committee was also joined intermittently by other professionals involved in various sciences and interests.

The primary purpose of the committee was to give advice and to help to resolve problems before they reached the final stages of planning and the public review process. Two environmental permits were required for the Tigrett project to go forward: the state 401 permit and a federal 404 permit. The Department of the Environment and Conservation administered the 401 permit. It addressed the limits on stream pollution and water quality in Tennessee. The Corps of Engineers office in Memphis administers the 404 permit, which is required when the filling of wetlands is involved. Unless the state approves the first permit, the corps as a matter of policy does not consider the 404 permit.

Therefore, the Tigrett Technical Committee could resolve conflicts with the requirements of these permits by pointing out the deficiencies in the proposals and correcting them. If the committee had no complaints, both permits stood a good chance of being approved, provided political intervention, interagency conflicts, and legal processes did not interfere. This way the permits could be approved quickly before funding disappeared or the political atmosphere changed.

The artificial swamps were the critical issue. If the committee decided that these swamps should remain intact, the objectives of the entire Tigrett plan and the principles of river restoration were in jeopardy. One of the underlying points about the artificial swamps was that obstructions (usually levees) created these swamps. Thus, the conclusion as to whether the swamps should stay or be removed should have been obvious: if the natural hydrology of the floodplain was inhibited by the levees that created these swamps, the river was not free to flow and the prospects for true floodplain restoration were remote.

Resolving the differences of opinions about these swamps had to be expeditious. Since changing or eliminating artificial swamps was a sensitive issue, the Tigrett Floodplain Restoration Plan did not highlight the subject, but we were unable to avoid it. And like the Nashville TWRA staff, some members of the committee did not see, or did not agree with, this reasoning. Some saw artificial swamps as beneficial for waterfowl, reptiles, fish, furbearers, birds, and a host of other wildlife species. This was true but only up to a point. The benefits of artificial swamps, with their low-quality habitat, were minimal at best.

"What does it matter," I suggested, "if you do find benefits from these swamps? The Memphis city landfill is the favorite site for Tennessee birders to find rare birds, even endangered species. Is it smart to create more garbage dumps for birders?"

Compromise on this point was a sticky issue with me. Committees tend to follow the trend of compromising first and evaluating the facts later. My rule of thumb was that there was room for compromise on anything but principles. Otherwise, the project should be scrapped or started anew. In this case, leaving artificial swamps intact violated all of the ecological principles in the Tigrett plan. At another location, perhaps, the swamps could have been justified—but not at Tigrett.

The status of the swamps continued to be the most worrisome issue for the duration of the project. The decision on whether or not these swamps had value should have rested mainly with the wetland managers, but it did not. I could not convince everyone on the committee that eliminating artificial swamps should not be open to compromise. At the committee's urging, the second draft of the Tigrett plan left 20 percent of the artificial swamps intact. This meant that one-fifth of the floodplain would remain obstructed and unrestored. The thing that most stuck in my craw was now part of the plan.

The first 404 permit application was finally filed with the corps in the spring of 1990. The review process continued for the next several years. The modified plan ran into opposition even before it reached the general public.

The Tennessee Department of Environment and Conservation was first to respond. Reversing its earlier positions, it seemed now not to want any of the Tigrett swamps drained. In this view, Tigrett WMA was already a model wildlife area, and channelization could be called the preferred management since it was the perfect method to create such swamps. It made no sense to me, and I was convinced that there had been political meddling. There were probably two main reasons behind TDEC's stand on the swamps. One was that the federal government had just recently supported a policy of "no net loss of wetlands," which often led to mistaken notions that no wetlands should be lost.[1] The second objection arose from a fear that if the swamps were drained, they would become soybean fields. "I'd rather have a stagnant swamp," Chester McConnell, a field representative with the Wildlife Management Institute and an advisor to the committee, said, "than to have these bottoms cleared for soybean fields." I agreed, but the idea that eliminating artificial standing water would automatically mean turning land over to soybean cultivation seemed misplaced to me.

The first objection illustrated how a once-useful statement about preserving wetlands had turned into a dogma. Whether the state water quality laws differentiated between artificial or natural wetlands, I cannot say. For that matter, I am not sure whether they discerned any difference in the quality of wetlands. Simply because water stood on the floodplain, in my own opinion and that of the northwest territories managers, was no reason to call it valuable wetland. For wildlife managers, wetlands should to be in appropriate locations and exhibit a reasonable number of natural characteristics to qualify as "good." This is not to say that artificially ponded water is useless for managing wildlife: whether the pond is a green tree reservoir (GTR) for ducks, a swamp, or some other kind of ponded water, it must have a legitimate purpose and not interfere with the natural hydrology of major streams. As Aldo Leopold, one of America's best-known conservationists, famously stated, "A thing is right when it tends to preserve the integrity, stability and beauty of the biotic community. It is wrong when it tends otherwise."[2] Restoring the artificial swamps to a natural free-flowing condition was the only way to provide sustained integrity, stability, and beauty to the river floodplain. It was as simple as that.

This brings us to the second objection. Unknown to many of us, there were hidden reasons for a select few to argue for these swamps, and these reasons had to do with the WTTP lawsuit and the mitigation lands—a subject I will take up in detail in later pages. Suffice it to say here that although TDEC and most others were innocent of this subversion, their decisions had probably been influenced by it. TDEC's position on the swamps would eventually change again—but there were those whose opinions would never change, and they would have a detrimental impact on the ultimate outcome of this thirty-year campaign.

For the moment, however, the differences of opinion over the artificial swamps probably came into clearest focus in the case of the Annie Laurel James farm, located on the Forked Deer River near Alamo. Modification of the river had turned the floodplain on the James property into a swamp. The Basin Authority had drained the back side of this swamp with a lateral ditch, much like the lateral channel used for Stokes Creek, and that was the center of the issue. Chester McConnell, TDEC, and others opposed the Basin Authority's lateral channel but wanted to keep the swamp, itself a product of channelization. Proponents for the Basin Authority, on the other hand, rationalized that this segment of the river would always suffer the influence of channelization and thus had to be maintained artificially to relieve the farmer's flooding problems. Typically, draining the swamps for soybean production was the likely goal of the landowner.

As McConnell and others believed, artificial swamps with some fish and wildlife were better than a ditched and drained soybean field. Maybe they were, but both of these ideas conflicted with hydrology of the river. That made no difference to them, however. Restoration of the river was fine, they said, but "don't drain the swamp." They were trying to have it both ways.

I offered a solution: why not work on a policy to forbid levees and unauthorized ditching in the floodplain, then eliminate the artificial swamps, save natural swamps, and restore the natural hydrology of the river? That way the farmers would have the drainage they needed and the tremendous benefits lost by channelization might be regained. But once more the answer was: "Don't drain the artificial swamps."

Let me be clear: waterfowl managers in the northwest territories did not support the Basin Authority's channelization methods at the James project; draining native swamps or digging ditches was not compatible with natural hydrology and failed to address the root of the problem. But in this case, the swamp was artificial, and we proposed draining it with natural outlets, while eliminating the lateral ditch.

The Tigrett Technical Committee, which met on the James property in July 1990, could have become a formidable proponent for river restoration had they resolved this single issue, regardless of whether or not they sanctioned the Tigrett plan. But the committee's thinking was not clear on how rivers could be restored without eliminating artificial swamps. As McConnell later admitted, "So we opposed the James drainage project, but eventually some of us were members of the WTT Steering Committee. You know we voted unanimously in our support to restore the entire river, and the entire floodplain system." Even so, the committee's stand at the time hindered any possibility for them to support the elimination of the artificial swamps in the Tigrett plan. The disagreement was destined to continue—but only for a while longer.

Chester McConnell and the Swamps at Ghost River

Positions taken at the James project bedeviled the thinking of the technical committee, created diversions in the sound arguments for river restoration, and thwarted the principles and practices that the managers in my district held sacred and could no longer in good conscience violate. Chester McConnell probably kept alive the issues of the rivers' conservation more than anyone alive; he never gave up. His natural resource management background was solid, and he already had my respect. He was a former manager of Laurel Hill WMA and a small game biologist with the state. But even though we shared so many of the same opinions about river conservation, our common cause hit a snag when it came to the artificial swamp issue. It plagued us like a case of malaria, and we both regretted it.

Our determination to resolve the disagreement was prompted by Ellen Danke, a reporter with the *Nashville Tennessean*, when she announced that she intended to do a feature story on the West Tennessee Tributaries Project. Since Chester and I would be interviewed, we knew that despite Danke's ethical and conscientious reporting, a story on the subject could easily highlight our differences of opinion, and this could confuse rather than enlighten the public. With this in mind, we pledged a concerted effort to lay this demon to rest before Danke's interview. We scheduled a meeting at my place in Dyersburg for November 17, 1991—the occasion for the final debate. The strategy was: "You show me yours, and I'll show you mine." Nothing was barred.

The process began with a daylong field trip at Tigrett. We spent the entire day discussing the consequences of the dead and dying swamps. At the end of the day, Chester thought he made his points fairly. "Well, how do you feel about it now?" he asked at the end of the trip. "That you weren't paying much attention," I replied. The problem was still with us. Heated discussions continued into the evening at my house until about 8 p.m. Finally, at 8:01 p.m., we miraculously reached agreement: the subject of rivers was off limits for the remainder of the evening. I uncorked a bottle of Scotch while Chester had a cold beer. We then had an enjoyable evening discussing nearly everything under the moon—except the rivers—until around midnight.

Our canoe launched the next morning in the Wolf River at the Bateman Bridge, near LaGrange, a few miles east of Memphis. Chester's example for the ideal attributes of an artificial swamp was a stretch known as "Ghost River." The river had been partially dammed by an accumulation of heavy sand known as a valley plug, likely caused by a poorly managed watershed. The swamp backed two or three miles upstream as a perennial pond. Unlike most, this river ran through the open pond at the lower end.

Portions of the river were exceptionally scenic. The running river meandered through the swamp beneath a green forest canopy. Groves of young cypress flanked both sides of the canoe on the lower third of the swamp, and a nice flow in the middle of the swamp uplifted and refreshed the entire scene. It was deceiving; dead snags scattered throughout the cypress told another story. Old stands of oaks and other dead or dying hardwoods had succumbed to summer flooding. Dying trees are so common in artificial swamps that people generally think the scene is natural. Chester made no mention of the dead snags. He had not yet conceded that without fresh oxygen all of these trees would eventually die, even the cypress. But I was there to remind him. Obviously, this floodplain had dried at one time, probably for more than one season, since the trees had to sprout and grow tall enough not to be drowned by the next high water.

Forested floodplains are possible only with seasonal drying. With disruptions to natural hydrology, however, little or no undergrowth can be found because the regeneration suffocates from prolonged flooding. This was the case in the permanently flooded section of Ghost River; here were some living cypress groves, a few straggling swamp tupelos, and dead or dying oaks. Only the young trees appeared to thrive, which is common in newly established swamps. The trees, as they grow older, will demand more oxygen. Without seasonal drying, these trees will also become dead snags. There are exceptions, but they are rare in Tennessee; and these exceptions do not preclude the need for a continuous supply of oxygen, often found where trees stand in flowing water.

At the end of our three-day session, Chester recommended that we let time and natural processes drain the artificial swamps on public land (especially at Tigrett WMA). It was a small compromise on Chester's part: he could accept natural change that corrected human interference, but in this case he did not care to see people correct their own errors. To me, of course, that sort of intervention was part of wise management.

Interestingly, although Chester had sued the Soil Conservation Service for its methods in promoting drainage, he himself had earlier promoted drainage in the main channel of the Wolf, working with others to develop a manual entitled *Stream Obstruction Removal Guidelines*.[3] I held back on pointing out the obvious: using the techniques he advocated in that manual would not only clean out the Wolf's main channel but drain the adjacent Ghost River swamp. I did not ask why his guidelines were not also appropriate to drain the old meanders and swamps at Tigrett. I knew that this was, for the moment, a dead subject. After his strong stand on the artificial swamp at the Annie Laurel James property, it would take Chester time and a little adjustment in thinking.

Chester's stream-obstruction removal guidelines were certainly good methods, and they were promoted heavily throughout the period of controversy involving the WTTP. But the guidelines did not mention the disposition of the anti-river, the artificial swamps, or the restoration of the natural river. Otherwise, they were very similar to our proposals for Tigrett, which was intended for all of the channelized rivers. Nevertheless, when the third 404 permit application for the Tigrett plan was filed, Chester's first word on his footnote was "Sorry." He recommended that the application be denied.

Chapter 12

The Drawdown Strategy
at Reelfoot Lake

In 1972 I returned to Tennessee from Washington, D.C., after being away for twenty-two years. It was late summer, and Reelfoot Lake was convulsing from the long months in the summer sun. The shallow backwater smelled of anaerobic gases, and there was an eerie stillness in the marsh that did not fit this paradisiacal place. The state of the lake seemed to indicate "hypereutrophication"—an extreme form of the process whereby dissolved nutrients in the lake become so abundant that they stimulate excessive plant growth that robs the lake of oxygen. With this "super-fertility" and stagnation, Reelfoot had reached, some said, its worst years since human interventions had isolated it from its parent, the Mississippi River. Complaints about invasive aquatic weeds and the terrible stench were common over the previous few years. A decline in the number and vigor of sport fish, rails, crayfish, pin minnows, furbearers, frogs, and other common wildlife species was also evident. A close look into the matter was long overdue.

Within a year, Reelfoot became one of my special projects in my new job with the state wildlife agency. From the very beginning, the lack of fluctuation in the seasonal water levels became a prime suspect for the lake's troubles. The flooding Mississippi River had taken care of this problem until the early 1900s when U.S. Army Corps of Engineers built the gigantic levees between the river and the lake. I had seen stagnant lakes before, and I could not remember one that was not caused by the hands of humans. When the biological vigor of these lakes declined, their recreational use also suffered. In 1973–74 I developed an experimental lake drawdown plan for Reelfoot to test the idea of varying the levels of the lake. It was not very successful. Like the natives of the Reelfoot area, some within TWRA considered a lake drawdown an odd recommendation, even though it was commonly done for one reason or another on artificial lakes the agency managed.

The upland game division and fisheries division of our agency historically did not join together to tackle biological issues. Ordinarily, they operated independently. To me, this unwritten rule was nonsense. A fisheries biologist had not been stationed at Reelfoot since Hudson Nichols in the 1950s and '60s, although the fishery was monitored by fishery biologists. The twenty-four-thousand-acre wetland was under the supervision of the wildlife area manager, a sort of general practitioner who was proficient in most wildlife and fishery matters.

For the best advice, however, I decided to consult Nichols, who had become the chief of fisheries for the state after his stint at Reelfoot. "What you say makes a lot of sense," he told me, "but what makes you think those Lakers would allow you to draw down Reelfoot Lake?" A tall, burly fellow hailing from Kentucky, Nichols had a special regard for Reelfoot Lake and the people living here. He had been a state fishery biologist for nearly thirty years. This experience added to his crusty, but distinguished, character. He had something to offer to young biologists, and I always took his counsel with gratitude.

The 1973–74 plan called for Reelfoot water levels to be dropped three feet to dry the shoreline of the lake. It was the first attempt to influence the ecology of the lake through the management of water-level regimes. The ideas for the plan came from several sources—my educational background in animal ecology, Nichols's experience, the experiences of the old-timers, similar situations at lakes in Illinois and Kentucky, and my own observations of Reelfoot. Everything pointed to the need for lake levels to fluctuate based on climatic influence. If a lake was somehow constrained from the influence of the climate, as Reelfoot was, it was logical to try to duplicate the effects by artificial means. Using artificial methods to lower the water levels of a wetland to expose the bottom to air and sunlight is called a "drawdown." Rather loosely, the term is sometimes applied to the natural drying of a wetland.

As in a river, the fluctuation of water levels in a lake is essential to its vitality. Droughty seasons and, conversely, high-water years are well known to produce dramatic changes in fish and wildlife production. During high water, the floodplains are available to fish and aquatic wildlife; when the water recedes, the floodplain is dried, refurbished, and made ready for the next flood event. A drawdown may be minor or extreme. "Extreme drawdown" is a relative expression, but it is generally used to describe lake levels lowered by more than 20 to 30 percent of their average depth, and sometimes such actions may even constitute total drying of the wetland.

I had observed an extreme drawdown while in graduate school at Southern Illinois University in Carbondale, Illinois. This particular lake was almost totally drained in 1962 to repair structures on it. Soon afterwards, I was able to observe the initial filling of Lake Barkley on the Cumberland River, an event similar to an extreme drawdown. If any data was collected on these projects, I have not seen it, but I did observe amazing biological responses on both lakes. Extraordinary fish spawns were evident the first year after these reservoirs were refilled. A tremendous number of minnows, crawfish, frogs, and other aquatic forms appeared. The biological richness and sport fishing were extraordinary for about ten years on both lakes. Hoping to achieve something similar at Reelfoot, my plan was simple: lower the level of the lake by two or three feet during the summer to dry the margins; this would be a minor drawdown. If the local citizens liked the results, a more exaggerated drawdown could be implemented later.

Nichols had tried new ideas at the lake before, and he was concerned that Reelfoot residents, ever vigilant about their wetlands, would object to the plan. Some years later Ed Hogg of Samburg told me that Nichols had initiated a brief winter drawdown in the 1960s to make repairs on the spillway. Apparently, it was also a good year to cut timber in the dried-up swamps. "He should not have done that," Ed said, "the timber should be left to grow for squirrels, woodpeckers, and the like. Besides, the first good freeze-up wiped out all of the bullfrogs that year." Nichols might have known the lake people better than I did, and he was not willing to risk his reputation on unproven management schemes. He stubbornly discouraged a summer drawdown, suggesting that I take another year to think about it.

But I believed that waiting was not an option. All anyone had done for Reelfoot during the previous three decades was wait. Such an approach may avoid controversy, but it can also allow more problems to accumulate. Nichols finally agreed to support a winter drawdown in 1975. While it would not serve all of the objectives, it was at least a start. Since a portion of Reelfoot was also a national wildlife refuge, cooperation from that area's manager was necessary. Through a 1942 lease agreement, a part of the U.S. Fish and Wildlife Service's duty was to operate the spillway gates that controlled the water levels of the lake. I discussed the plan with Wendell Crews, the NWR manager. "I reckon we can do that," he said. But he did so at some risk to his reputation as a cautious manager; the gates had been lowered according to a designated schedule for many years, and many expected it.

We attempted the drawdown the following winter, in January 1976, and tried again in 1978; the intent in both cases was to hold lake levels down by as much as three feet until April 1. Both drawdowns failed. No sooner had the lake levels started to fall than the rains began. The lake reverted to its original level and nothing dried out. From these attempts, we learned at least one valuable lesson: if natural conditions are unfavorable, wait. Fortunately, no one seemed to notice the attempted drawdowns or the results. In retrospect, a winter drawdown is a gamble from the outset. It all depends on the climate: although some winters are dry, they are impossible to predict. There is no good choice but to play the most favorable odds.

Winter drawdowns are a common technique used in northern lakes to freeze, kill, and control noxious aquatic vegetation, but they had a particular downside at Reelfoot. Local residents such as Ed Hogg who were familiar with natural drawdowns in the winter saw them as negative events resulting in freeze-ups that killed fish and reptiles. Since freeze-ups and the attendant wildlife mortality are natural occurrences in the ecology of wetlands, they are obviously beneficial in the long run. But many people do not regard such episodes that way, even when they arise from natural causes. Given such perceptions and the failure of our first attempts at a drawdown, I would not return to the subject until a few years later.

During the interim, Hudson Nichols retired. I took his earlier advice and waited until local sentiment appeared to be more favorable for conducting a summer drawdown. By the mid-1980s, with interest in improving Reelfoot on the rise, the timing seemed better. Or so I hoped.

The 1985 Drawdown Proposal and Its Justification

For years the administration of the lake had not been well defined. For the most part, the various agencies responsible for it worked independently of one another: Tennessee State Parks administered 270 acres; the Fish and Wildlife Service administered 8,000 acres for migratory birds; and TWRA administered some 20,000 acres as a wildlife management area. It was cooperative arrangement of shared tasks (one of which I have already mentioned: the Fish and Wildlife Service's operation of the flood gates for regulating lake levels). Unfortunately, prior to the 1980s, these various groups did not meet often enough to form a joint "think tank" for discussing the management of the lake as an ecosystem. But things finally began to take a turn for the better in 1982 with the appearance of a comprehensive report by Wintfred L. Smith and T. David Pitts of the University of Tennessee at Martin. Entitled "Reelfoot Lake: A Summary Report," it was the first paper of its kind, including or referencing most of the significant history and data known about the lake.[1] It pointed out several disturbing ecological trends, and before long, Reelfoot would be at the top of the list of projects needing attention from state wildlife and conservation agencies.

The General Assembly passed a resolution on February 10, 1983, "to create a special task force committee to study the management, preservation, and development of Reelfoot Lake." The Reelfoot Task Force was made up of members from three state agencies, two federal agencies, the University of Tennessee at Martin, and three local citizens, including two resort owners, Robert Gooch and Al Hamilton, and a Lake County historian, Betty Sumara. Afterwards, a TWRA committee chaired by Harold Hurst was designated to draft the "Reelfoot Management Report" (1983), which targeted seven categories of concern: soil erosion/sedimentation, water quality, shoreline management, water-level management, aquatic vegetation, public understanding of threats to the lake, and the state and federal management responsibilities at Reelfoot. Committees were then assigned to address each category.

It was a stellar effort, and lake managers everywhere would do well to reference all of these reports before developing or updating future work. These action committees moved at such a pace that the "Reelfoot Management Report" remained only a draft report. But though it was left unfinished, many of the recommendations were nevertheless implemented.

In 1984 the Tennessee Department of Health and Environment released another report entitled "A Clean Lake Study: The Upper Buck Basin of Reelfoot Lake, Tennessee." The report identified in great detail the parameters that influenced the lake's water quality,[2] and it was an excellent supplement to the work done by the UTM professors Smith and Pitts. Interest continued to grow to "do good things" for the lake.

Professor Smith, along with his UTM colleagues Andrew Sliger and Wesley Henson, saw the lake as a special project. These researchers would conduct biological, chemical, and physical studies at the lake over the course of their entire careers, probably much of it on their own time. Much of this research was practical as well as academic, and they helped the lake managers immeasurably by sharing their findings and readily discussing all aspects of the lake's ecology and management needs.

From the standpoint of practical management, it became increasingly clear that a drawdown was one of the few options left to improve the lake. Other field biologists and managers who spent their careers managing wetlands and inland freshwater lakes supported this view. At Reelfoot, the time for an extreme lake drawdown was long overdue. Little did I know that this seemingly innocuous proposal would upset the foundations of the local society, pit agency against agency, neighbor against neighbor, and state against state. Who could guess that it would cause commotions much like those created by the West Tennessee Tributaries Project?

The 1973 drawdown had been an in-house project not fully scrutinized by the public, the conservation and natural resource agencies, or other professional interests. In fact, there was no interest, although the indicators for the future of the lake were obvious to anyone willing to observe. The 1985 drawdown proposal, on the other hand, would be scrutinized beyond our wildest imaginations.

Oddly, the biological problems at Reelfoot were quite simple. It took very little to see that these wetlands were isolated from the river that created and sustained them. It had become a very large pond, far different from the native lake it had once been. Options for managing Reelfoot had been reduced: neither the levees that separated the wetland from the river nor the spillway that regulated its water levels would ever be removed. The choice was to do nothing or to resort to artificial management, and for us, doing nothing was not an option.

Even though the marshy lake was still considered a natural treasure by the state, it could never regain its former beauty and vigor as a natural wetland. Wise management, however, could certainly improve it. Nevertheless, the prevailing opinion was to leave Reelfoot alone. Doubtless no native of the area would harm a single lily on the lake if they really thought it would do harm; no one had a greater concern for Reelfoot than the local folk. Yet, for whatever reason, very few took the time to get opinions from the lake managers. Rather, they seemed to prefer hearsay and limited experience, or they simply rejected science. They seemed to think that the lake was theirs alone to understand. I can sympathize with such feelings to a degree, but I believe that the local citizens still had a responsibility to listen to possible options that could benefit the lake.

Isolated from the river that had formed it, Reelfoot Lake no longer had "room to breathe." Gone were the cycles of natural flooding and drying and the exchange of nutrients, fish, and other aquatic organisms needed for the lake's vitality. Saving the lake depended solely on the ability of its managers to mimic what nature had done. While initiating a drawdown and regulating water levels were not the preferred management, these were the best options we knew. But not many agreed with this assessment in the 1980s.

We went to great lengths to communicate our reasoning for the new proposals. Monthly ad hoc sessions were held with the Reelfoot Technical Committee, which I chaired. The members of the committee, like that of the later Tigrett advisory group (see chapters 10 and 11), were environmental regulators, reputable scientists, and both opponents and proponents of the plans. The committee gave advice, helped resolve many misconceptions, and clarified the purpose of our work in addition to making many useful recommendations.

The TWRA produced an especially informative video featuring the well-known professional fisherman and television personality Bill Dance, who supported the drawdown: "I cannot stand to see the Reelfoot Lake lose the excellent fishing I have known." His feelings summed up the opinions of the silent majority—those who cared and knew the history of Reelfoot Lake. The video was distributed widely to local television stations and shown at numerous public meetings.[3]

Most of the complaints about fishing at Reelfoot concerned water quality, the filling of the lake with sediments, and unfavorable changes in the fish populations. The extreme lake drawdown promised to address all of these concerns. Lake depths, in particular, were the major factor in the health and vitality of the fishery. Original depths of up to forty feet had once been recorded for Reelfoot, but current depths of more than fifteen to eighteen feet were rare. More than five thousand acres of the lake were eighteen inches or less in depth, which made them highly unfavorable as fish habitat, and sediment deposition of only a few inches could only worsen the conditions. Thus, anything that could increase the depth would be good for fish.

The answer was clear: we had to either shrink the substrate and the bottom of the lake, or increase water levels. A few inches of additional water could immediately increase the depth of the lake of the marginally shallow waters and improve the fishery of the entire lake. The drawdown plan proposed both of these options: first, we would dry the bottom of the lake—the unconsolidated muck—and, after the drying, we would raise the water levels a foot higher. It would not only improve the fishery but also provide better boat access across the lake.

The drawdown would also help to control the noxious aquatic weeds that were the source of many complaints by lake users. Visiting scientists such as E. O. Gersbacker and E. M. Norton were concerned as early as the 1930s that vegetation was spreading rapidly and that the north end of the lake would soon be filled beyond the possibility of commercial use.[4] And in another research study, conducted more than forty years later, Wesley Henson and Andrew Sliger stated, "In 1983 many areas of Reelfoot Lake were clogged with submersed vegetation, particularly curlyleaf pondweed (*Potamogeton crispus*) and coontail (*Ceratophyllum demersum*)."[5]

Aquatic weeds like the ones these scientists described respond to water depths. Waters of less than three feet in depth are certain to encourage these weeds. Soil characteristics on the lake bottom also make a difference in weed growth. In conducting a drawdown, we hoped to dry and change the soil composition sufficiently to make the lake bottom unfavorable for many of these invading plants; allowing it to dry for more than one season could eradicate some of the hardiest weeds. Consolidation of the soils through the drying process could also shrink these soils and add to the depth of the lake, which in turn could promote more desirable aquatic plants.

Fish population samples taken since 1952 showed that composition and weights of sport fish in the lake were gradually in decline while commercial or rough fish increased. The percentage of game fish (crappie, largemouth bass, yellow bass, bluegill, and other sunfish) in our samples declined from 62.8 percent in 1952 to 10.2 percent in 1985. Rough fish (carp, buffalo, gar, catfish, bowfin, shad, and others) increased from 37.2 percent to 89.8 percent for the same period. The average weight of crappie harvested by sport fishermen dropped from 9.5 ounces in 1955 to about 4.0 ounces from the early 1960s through the 1980s.[6]

So it appeared that within the period between 1952 through the 1980s, the weights of crappie continued to fall with or without commercial fishing, and the composition of game fish were also less. The strongest conclusion by the biologists was that the lake itself was changing. To them this was an indication that high nutrient levels and low oxygen content in the lake were more unfavorable for sport fish. Samples at the bottom of the lake also showed very low oxygen levels—sometimes two parts per million, sometimes as low as zero. This was a critical finding when six parts per million of oxygen is considered necessary for viable fish populations. This oxygen deficit resulted mainly from a widespread accumulation of organic matter, which consumes oxygen required by aquatic organisms such as fish and certain insects. Sometimes this accumulation could be more than five or six feet thick. This silt-like sediment not only contributed to low oxygen conditions, it also resulted in poor spawning habitat for fish.

The fish barely survived these spawning limitations. They were limited to laying eggs on logs, sandy bottoms, and other places near the surface where the oxygen levels were better, sometimes more than twelve parts per million. Good fishermen knew this fact and took advantage of it by fishing only these structures for bass, bluegill, and crappie during spawning season.

The low weight of fish and their decreasing numbers became a major part of any effort to effect the drawdown. Contributions from the Mississippi—fish, food, surging water, and a host of other essentials—still occur at Reelfoot but not often enough. The Mississippi is truly the mother of Reelfoot, just as rivers are to all native wetlands in the region. For wetlands to function in a natural, sustainable way, this relationship had to be a permanent. When it ceases, wetlands begin a trend toward the highly fertile conditions we see at Reelfoot Lake—hypereutrophication. Only skillful management can inhibit this trend and prolong the lake's decline or extermination. The extreme lake drawdown was one management technique that sought to address these problems.

The Hypereutrophic Lake

Landlocked wetlands like Reelfoot and the artificial swamps along West Tennessee's rivers are often extremely fertile. Explaining that Reelfoot was a hypereutrophic lake was not a good way to hold the attention of an audience. From the beginning, it seemed contradictory that a super-fertile lake would *not* produce the abundance and variety of fish and wildlife that everyone wanted. But overfertilize your garden and see what happens: the fertilizer will "burn up" or kill your plants. Very few would deny that the marshes of Reelfoot smelled rotten from July through October. Aquatic animal life in these marsh conditions is low, and there is little successful fishing. These were clues to a hypereutrophic condition.

The kinds of sediments that accumulate in a wetland have much to do with its fertility. Sedimentation of a wetland is common and part of a natural aging process. Over time wetlands naturally fill with sediments and become dry enough to support trees. Two major categories of sediments have contributed to the filling of Reelfoot Lake: *inorganic sediments* and *organic sediments.*

The accumulation of inorganic sediments, such as sand or clay, is the result of erosion. Unless these soils carry foreign chemicals (such as pesticides or nutrients like fertilizer), the most detrimental effect is the sediment build-up itself. TDEC scientists estimate that sediments fill Reelfoot at an average rate of one foot every thirty years, most of it from soil erosion; this condition is twenty to thirty times faster than desired or expected from natural erosion.

Organic sediments—dead plants and animals—mostly originate within the lake and provide nutrients to it in much the way that common commercial fertilizers provide nutrients to cropland or gardens. These deposits are natural in healthy wetlands and become problematic only when the deposition is excessive and inhibits the production of available oxygen to aquatic life.

There was hardly anyone who could remember years when Reelfoot dried out enough to allow one to walk through its foul-smelling marshes. Thus, few knew that these marshes would not smell so bad if the lake had experienced periodic drying. In a healthy lake, this process is natural; indeed, it is necessary to help correct some of the lake's oxygen problems.

During the 1800s, no doubt, the water levels of Reelfoot Lake rose much higher, but the lake also fell to much lower levels during droughts. In fact, Bennett Johnson, Lewis Michael, and other long-time residents of the lake could remember a period during the 1920s when they were able to walk across the north basin of the lake on dry soil. "You didn't wade in knee-deep mud after that drought," Michael said. "The bottom of the lake that dried was as hard as the dry ground for years after the lake was refilled."[7]

So, it was not uncommon for the lake (or any natural wetland) to fall lower than normal levels during dry seasons. These periods of extreme drying work in concert with the climate and the natural rise and fall of rivers, and they are key factors that affect the vigor and longevity of wetlands like Reelfoot Lake.

The purpose of the extreme drawdown was to replicate the normal drying that had occurred during Reelfoot's natural history: droughts could do the job, or humans could mimic it. In a relatively short time, the entire mess of exposed organic sediment dries, begins to decompose, and turns into soil. It can be fit for a garden when exposed for an entire summer or two and gives new meaning to the words "wetland management." Wetlands so managed pose no threat to the ecology of their floodplains. An extreme drought or a drawdown allows the bottom of the wetland to become useable rather quickly for the plants and animals endemic to the place. It could work not just for Reelfoot Lake but also for artificial swamps and other wetlands isolated on the floodplain.

Reelfoot has characteristics of both a natural and an artificial wetland—natural because it is an oxbow, artificial because it is trapped by levees that preclude the effects of natural hydrology. Like Reelfoot, Tigrett and other stretches of the rivers offer classic examples of hypereutrophic wetlands because they have been artificially modified to become ponds instead of naturally drained floodplains.

The Impact of Levees

Along with channelization, levees are a major cause of wetland degradation. Such structures were constructed in the Reelfoot vicinity after the formation, around 1909, of the Lake County Levee District. Working with its counterpart in Kentucky, the organization built a levee from Hickman, Kentucky, south along the east banks of the Mississippi River to the high land at Tiptonville.[8] This was followed in 1917 by the construction of a levee crossing the drainage at the south end of Reelfoot Lake. The intent, once again, was to exert total control over nature, flooding, and the water levels of the lake. With the addition of a spillway, authorized and funded in 1929, Reelfoot would be controlled and maintained "at a constant and uniform level."[9]

Levees built in the floodplain, as we saw along the Obion and Forked Deer as well as at Reelfoot, affect all components of the wetlands. Not only do these structures prevent the natural rise and fall of floodwaters, but they also prevent the exchange of nutrients and biota between the river and the wetland. This exchange between ecosystems allows an interchange of many species of animals and plants—fish, freshwater eels, turtles, crustaceans, microscopic plants and animals, and other forms of aquatic life—important to the ecological equilibrium of both systems. When the exchange of water is restricted, the desired health, natural beauty, and uses of the lake are also compromised. Food chains and important natural processes are disrupted or absent. Eliminate the crayfish, for example, and the food chain is broken. Fish, birds, raccoons, and other animals that depend on crayfish for food may not survive. Levees also eliminate large areas of wetland habitat. Without the historic normal high stages at Reelfoot, the lake could not reach its full potential.

In a healthy lake, as Reelfoot once was, rising water spreads beyond the normal level of the lake well into the floodplain. Such high lake stages are mostly seasonal and relatively unpredictable events. But when this surge of fresh water occurs, insects, earthworms, mollusks, and other sources of food become readily available to fish, waterfowl, furbearers, and other wetland species. During the spring and early summer, the inundated shores provide excellent spawning grounds for fish and brooding areas for waterfowl and mammals. Thus, the energy of the lake receives a welcome boost, which is passed on through the food chain: crawfish eat insects, fish eat the crawfish, eagles eat the fish, and so on. Populations expand and their general health improves. Numerous physical, biological, and chemical changes occur to maintain the lake in a healthy state. Thus, the lake also becomes more attractive for outdoor recreation. The effects of high water may benefit the floodplain for three or four years after the flooding. Thus reenergized, the lake is generally sustained until the next flood event.

Since only a few users of Reelfoot could remember the effects of high and low water, their importance was little understood and appreciated there. The cypress trees at Reelfoot offered a telling example of what can happen when the natural fluctuation of water levels is interrupted. When stranded in permanent water, these trees live at the very edge of death. But in a natural flooding-and-drying regime, the trees that survive can be expected to live vigorously, just as nature intended. This principle can be mimicked to a large degree by an extreme lake drawdown.

In fact, we saw that many of the trees living in marginal conditions around the lake would benefit from the drawdown, particularly those in shallow water. Tree growth improves when the roots are exposed to drying. Cypress, like other trees, do not sprout except on dry ground, and they cannot grow as expected unless the soil is drained and their roots are exposed often enough to provide oxygen during the growing season. Understanding this process has made it easy to conclude that Reelfoot had once had a basin sufficiently dry for trees to sprout.

The swollen buttresses on cypress trees standing in Reelfoot's open areas have long been a mystery to many, but I believe them to be indicators of the stress caused to trees by stable water levels and oxygen depravation. The major function of tree roots is to provide oxygen and nutrients to the organism, and the large buttresses seen on Reelfoot cypresses represent, in my opinion, an adaptation by the trees in response to their depravations. Although this supposition, to my knowledge, has not been scientifically demonstrated, we do know that the buttresses occur at the ordinary pool elevation of the lake, precisely where most of the dissolved oxygen is found. This condition in cypress can be seen not only throughout Reelfoot but also in swamps that rarely, if ever, drain. These are "false buttresses" with numerous small rootlets, and we find them only on trees that have survived long periods of standing in relatively stable water. Such trees apparently adapt in the most expedient way to survive as long as possible—although they will eventually die from stress.

We have learned that when stagnant pools are drained, the swollen buttresses are left high and dry at exactly the common water level of the pools. In visits to Louisiana lakes that had been subjected to extreme drawdowns, I saw remarkable evidence that the trees could recover. The lake levels at Reelfoot have never fallen enough to demonstrate a similar transformation, although we have examined the trees and know their true condition. The bole (that is, the stem or trunk) of the cypress looks quite normal beneath the false buttress, all the way down to the original roots.

Managing the Stages of Natural Wetland Succession

Thus, levees or dams greatly affect the hydrology of a floodplain and all living things endemic to it. But what constitutes the "desirable" conditions of wetlands is generally a matter of judgment. Once the objective of wetland preservation is determined, managers often interrupt ecological succession to reach the objective. The general rule of thumb is that, however the wetland is managed, it should look and function in a way very similar to that of a natural wetland. Understanding the influence of environmental factors common to the region—climate, topography, soils, age, and the like—a manager may use controlled fires, resource harvesting, flooding, drainage, or other means to meet the objectives for wildlife and the public while maintaining the historic use and ecological characteristics of the wetland.

Physically, ecologically, and in other ways, wetlands go through dramatic changes as they grow older. A river wetland like Reelfoot Lake generally evolves through several stages. When it reaches its older stages, shallow swamps, marshes, glades, and even meadows predominate in some sections. The final stage will tend to be a mature bottomland forest, as has already occurred on Grassy Island and other margins of Reelfoot. The composition and populations of plant and animal communities change with the wetland. A lake may have, for example, an enormous population of frogs during the first few years after it is created, with few creel-size bass. After a decade, the lake is likely to reverse and have very few frogs but many creel-size bass. However, as the lake reaches the final stage, the frog-bass population ratio is likely to change again. Maintaining any particular population level of frogs or bass is a matter of maintaining the habitat preferred for the population desired.

Reelfoot Lake at the time of the 1985 drawdown showed characteristics of approximately the last quarter stage of an aging lake. With its natural succession interrupted by levees and other factors, the biological age of the lake was probably accelerated by as much as fifty to seventy-five years. Whether or not the targeted objective of preserving Reelfoot at its current stage can be done depends largely on a good management plan and support from the public. A plan for the desired future of Reelfoot had not been developed until 1985. The plans we designed then are still valid but have yet to be implemented as of this writing.

In 1985 our judgment—for both the immediate drawdown and the Reelfoot Lake fifty-year plan—was that simulating droughts by a drawdown and then raising water levels were among the few things we could do to set back the aging process of the lake and thereby provide and maintain the quality of habitat necessary to meet the plan's objectives. This assumption had been tested to our satisfaction numerous times before by others as well as by our own trials. We had observed numerous lakes in Florida, Louisiana, and Arkansas, in addition to lakes in Illinois and Kentucky, and tested smaller wetlands in Tennessee. Furthermore, we spent considerable time discussing extreme lake drawdowns with wetland managers who had years of experience with this technique.

Crucial to all our management recommendations was the principle that there is an inseparable relationship between wetlands and the native rivers that produced them. Thus, whatever a lake's origin—whether it was once natural or began as an artificial wetland—the principles of the natural wetland ecosystem became the key to the desired health and vitality of the wetland. This underlying tenet makes it impossible to think about lakes, swamps, and marshes and *not* think about the rivers that created them. Keeping this in mind, we had no reservations in recommending that an extreme lake drawdown was one of the best alternatives for addressing the highly eutrophic conditions at Reelfoot.

Chapter 13

Implementing the
1985 Extreme Lake Drawdown

The objectives of our plan for an extreme drawdown at Reelfoot went considerably beyond those of the winter drawdown Hudson Nichols and I had proposed a decade earlier. With this one, we intended to dry as much of the lake as possible for as long as possible, thus producing a simulated drought. We would follow this by raising the lake higher upon refill, thus simulating seasonal high water. During the interim between drawdowns, we planned to manage water levels at random stages to mimic the water levels one sees during average years on a natural lake: after all, climatic influences on the wetland environment vary from season to season, and no two seasons are the same. (Again, nature abhors stability.) After making a few basic diagrams to supplement our plans, I took the package to Harold Hurst, the Region 1 manager, in the fall of 1984. I could not have asked for a more supportive boss: if the logic was right and Harold was convinced, his support, commitment, and loyalty were unshakable. He remained our closest ally until he retired from the agency in 1999.

Assembling a Team, Getting Advice

Meanwhile, Paul Brown had arrived in January 1984 to fill the vacant manager's position at Reelfoot. Paul had all the credentials needed for the job. He was from Paris, Tennessee, and a graduate of the University of Tennessee at Knoxville. He had also done graduate work in fish and wildlife at Tennessee Technological University. He came to Reelfoot from the position of wildlife officer for Jackson County. Within a fairly short time, he tied up the loose ends and corralled the rapidly developing chaos. Voluminous amounts of technical material and the history of the area had to be learned quickly to cope with rising interest and controversy at Reelfoot. Getting to know the people—something that could have been his most difficult challenge—proved to be easy for Paul.

The state's new chief of fisheries, Wayne Pollack, was more than willing to come to the regional office to discuss the new drawdown proposals. Harold, Paul, and I met with Pollack and his fisheries biologist, Robb Todd, in June 1984. Pollack heard the same presentation we had earlier given to Nichols and Hurst, which took about fifteen minutes.

Paul Brown, TWRA wildlife manager at Reelfoot Lake, directed
projects there that often set standards for managing wetlands
along the Obion–Forked Deer rivers. (Photo by Jim Johnson.)

Afterwards, he recommended that we contact the fisheries division in Florida and ask whether they would send their experts on lake drawdowns to Reelfoot. Florida's Freshwater Fish Division had been involved in lake drawdown techniques for several years. Forest Ware, the chief of the Florida fisheries section, did not hesitate to comply with our request and came himself. Ware's visit was short. He immediately saw the similarity of Reelfoot's biological status to that of lakes he had worked with in Florida. A week or two after Ware's visit, in July, two of his crack fishery biologists—"Bucky" Wegner and Vince Williams—showed up at the lake and met us at the Reelfoot office.

These biologists were not simply doing a job; they were submerged in it. Time and distance meant nothing to them when dealing with the science and management of warm-water lakes. The only amenity they required was a barbecue cookout, which suited them better if a good bottle of corn whiskey was included with the hors d'oeuvres. This entertainment had to wait, however, because they wanted to get onto the lake. So did we. A field trip was conducted within an hour after they arrived.

The field trip took less than half a day. Wegner and Williams both believed that our assumptions about Reelfoot were correct—that the lake required periodic drying and flooding to maintain the relative health and vigor of its aquatic life. They agreed that the bottom of the lake was essentially an "anaerobic soup" and were not surprised that we found oxygen deficits in the lake. They felt that the oxygen content in fish spawning habitat was, in all probability, extremely low and inhibited fish production. They also agreed that the fishery at crucial times of the year was probably on the brink of an oxygen crisis and that it probably reached that level often. It was comforting to know that their independent conclusions were about the same as ours.

They went into some depth to explain their reasoning and elaborated on the basic science behind the drawdown technique. The metabolism of the lake was highest during hot summer months, they told us. Oxygen production from aquatic microscopic plants reached its zenith on sunny days, but while these tiny plants supplied a lot of available oxygen, the demand for it—from fish and, indeed, the entire aquatic ecosystem—was also extremely high. This brought us to the critical part. A few cloudy days during the hot summer months could easily limit or cut off photosynthesis conducted by the plants. Since this also reduced the main supply of available oxygen, the supply would plummet. These astute biologists were puzzled, in fact, about why extreme fish mortality at the lake had not already occurred and how the lake was able to maintain the current level of sport fishing. We wondered, too, and saw it as one more sign of Reelfoot's unique character as an earthquake lake. We speculated that the sunken forest that had contributed to the lake's super-richness (and now the build-up of organic muck) had also given it a tremendous energy bank and great habitat diversity, and for years aquatic life had managed to survive. Now, however, the delicate balance between conditions favorable to such life and the conditions that could kill it was threatening to tip in the wrong direction.

Wegner pointed out that a three-foot drawdown along the lines of what had been attempted years earlier would hardly resolve the problem. Since the lake had accumulated such a deep layer of organic sediments, up to six or eight feet in some places, Wegner emphasized that a drawdown of at least six feet was needed and that, in fact, an eight-foot drawdown would be far better. He said that the lake should dry out as much as possible and that any dense weed growth, which was likely to happen after the drawdown, would have to be eradicated before the lake was refilled. Otherwise, invasive giant cutgrass and black willows would have resort owners around the lake up in arms. We had considered this and certainly agreed with Wegner's advice.

The biologists emphasized that the benefits from the drawdown would accrue with every foot the lake was lowered. Since the lake was shallow, hundreds of acres of mudflats would be exposed. In fact, they recommended that the drawdown be designed to expose at least 50 percent of the lake's shoreline. They acknowledged that some fish would be trapped in shallow basins and would probably die but that this should not alarm us, since more than an ample population of fish and other aquatic life would survive to refurbish the lake. The argument made sense because such things happen in nature during drought years. A major die-off could actually benefit the lake, thinning out the fish populations as it improved the habitat. The aftereffect could well be a large amount of favorite fish habitat. This in turn could greatly increase the growth of fish and create an enormous flush in fish production on which predator fish, such as largemouth bass, depended. We fully expected this combination of events to lead to a more desirable ratio of game fish to forage and rough fish. A major point was that, once refilled, the lake would have a bottom far more favorable for fish to spawn. It would also be a better habitat for those organisms that served as fish food. All in all, the processes were similar to those expected in natural lakes, and this was the result we hoped to see. It was easy to see, too, that the same principles also applied to artificial swamps along the Obion and Forked Deer rivers. Before the Florida biologists left, they advised us that Louisiana also used the drawdown technique for the primary purpose of controlling noxious aquatic weeds.

In the final Reelfoot Lake drawdown plan, the lake would be lowered by only 5.8 feet at most since this was the maximum capability of the spillway. The exposed lake bottom was scheduled to dry for a minimum of 120 days based on recommendations from the Florida biologists. The drying period would extend from mid-June to mid-October, because evaporation and drying ordinarily were greatest through the hot summer and dry fall months. This schedule also minimized conflicts with outdoor recreation on the lake because this was a period of very low use. On refill, the lake was to rise to an elevation of 283.0 feet above mean sea level, or nearly a foot above normal pool, before the spillway gates would be opened. If the weather cooperated and heavy rains did not thwart the plan, this schedule would mimic natural cycles and provide maximum benefits.

A special "scoping" meeting was scheduled for August 15, 1984, at the Peabody Hotel in Memphis. Twenty or so scientists, technical people from various related fields, natural resource managers, and several others came to this large affair. The purpose of this quasi-scientific gathering from all across the United States was to assess the pros and cons of the proposed Reelfoot Lake drawdown as part of the environmental impact statement (EIS) process. The simple plan seemed strange to some of the attendees, whose specialties did not include experience with lake drawdowns or, for that matter, wetland management. Nevertheless, they were invited because their fields of work somehow related to the federal involvement in this project.

Because the Reelfoot drawdown plan survived the scrutiny of such meetings intact, we knew that we had either proposed something so out of focus that no one yet understood what we were talking about, or that we had proposed something worth considering: tomfoolery or risk-taking proposals simply do not endure such critical reviews. But a looming public relations problem remained. Regardless of what scientists thought about the drawdown, their conclusions had no standing with the people around Reelfoot; science seemed unable to resolve any of the local anxieties about "draining our lake."

Public Resistance and Leadership

As the scheduled date of the drawdown, May 27, 1985, approached, a number of factors—events, personalities and leadership issues—converged in a way that would affect the outcome.

Hurst and Myers supported us on the Obion and Forked Deer rivers projects, but the circumstances and long neglect of Reelfoot demanded a special interest from others. Tim Broadbent, the Region 1 fisheries biologist, and Wayne Pollack were essential advisors. Strangely, the Reelfoot drawdown never attracted much scientific curiosity from our regional or statewide peers, so we were more than pleased to have Broadbent and Pollack's involvement. Otherwise, interest in native wetlands seemed to be peculiarly limited to our small district within the region. Reaction from the public was similar. Wetlands seemed to create interest only when their disappearance seemed imminent. At this point, public reactions and public relations could change, sometimes disastrously, overnight. Without leadership, such shifts were often to everyone's disadvantage.

Gary T. Myers, TWRA director, supported many of the field managers' recommendations for river restoration in West Tennessee before politics and legal squabbling intervened. (Photo courtesy of TWRA.)

Harold Hurst, TWRA Region 1 manager, supported restoration projects that signaled important changes for waterfowl-wetlands management at Reelfoot Lake and the rivers. (Photo by Kathy Krone.)

For about four decades, starting in the 1930s, local leadership at Reelfoot Lake came from Marvin Hayes, who was known as the "Reelfoot Godfather." A one-time state representative, he tended to support the interests of his commercial fishing business and also defied his competition with considerable success. Like it or not, Hayes's leadership was effective, but his era passed. Local leadership suffered until Rich and Suzy Cooper arrived. During the 1970s, these resort owners were effective representatives for the people at Reelfoot, and Rich presented their views to the TWRA commission on a regular basis. Sometimes they seemed to be a lone voice, as other business owners and officials around the lake generally kept silent. Nevertheless, the Coopers' skillful and energetic efforts helped both the people and the TWRA commission on worthwhile projects around the lake. Unfortunately, they sold their business prior to 1985 and left the state, with no one to take their place. This leadership void would hurt us.

During the summer of 1984 we had less than a year to make final preparations for the implementation of the drawdown. A year might seem like a lot of time to prepare for opening the spillway gates and dropping the level of the lake, but there was more to it than that. Monitoring techniques had to be in place, research needed to be done, and the public's concerns (and discontents) had to be addressed. Funding and planning were needed for aerial expeditions that would allow us to photograph and map the lake. Environmental critics wanted their questions answered. Several construction projects, such as those involving boat access to the lake at the low-water stages, required identification and design. Navigation and circulation channels to the lake's four major basins had to be addressed. The list of things to do seemed endless.

No one living had seen the lake, vigorously controlled by the artificial spillway, at more than three or four feet below its normal pool. Drawing down the lake would give us a chance to make repairs on its facilities and to note its exposed internal structure for boating and fishing. Since the sunken cypress forest over the years had been sheared near or at the surface of the lake, the underwater trees—some of which had not been seen for more than 150 years—were hazardous to boats. With the drawdown, boat trails could be marked and cleared through the most hazardous areas. Accurate maps could be sketched showing the location of these stump fields as well as fish structures and the best navigation routes in the lake. Boardwalks, hiking trails, fishing piers, boathouses, and docks could also be repaired.

However, before we barely had the chance to point out these opportunities, a public outcry arose: "TWRA is planning to *drain* Reelfoot Lake!" The words caught on like wildfire. Opponents refused to acknowledge the term "drawdown": to them this meant draining the lake bone-dry—even though that would probably be impossible. Nothing we said could change this opinion. Public meetings and an environmental assessment addressing this issue did not help. Without strong, credible leadership among local residents, generally speaking, very little can be accomplished.

Meanwhile, a resident owner of a local lodge—I'll call him "Mr. Legend"—had begun several ongoing initiatives involving the lake. Starting around 1983, his actions created a stir among the politicians and many local people. Mr. Legend was intense and often very convincing about the things he believed in. The lake was his life, as it was for most of his family. His enthusiasm was potentially beneficial for the future of the lake, and we needed his support. Yet, he had little, if any, interest in the biological issues that we had to address in order to improve Reelfoot's health. Instead, he concerned himself with the administrative procedures for the lake's management, the laws that governed the lake, and questions about its ownership. With great energy, he researched the history of the lake and found what he considered to be several gaps in the law and administrative authority over the lake. One of these involved the issue of riparian rights. Mr. Legend believed that the state had not purchased all of the land beneath the lake and that certain landowners—possibly his family—still possessed some of these rights, which, he argued, our plans would disturb unfairly.

The riparian issue alone was not enough to satisfy Legend's boundless research: the more he discovered, the more he searched. He became interested in every aspect of Reelfoot Lake administration—except the drawdown. Legend began a one-man campaign for numerous projects around the lake—some of them beneficial, some not. Even as his initiatives gained the support of a small group of local people, he relied on politicians to champion his interests In this regard, he did very well, and his efforts resulted in two lawsuits. The first was over the riparian rights issue. The second concerned TWRA's alleged efforts to eliminate duck blinds. While the lawsuits did not specifically target the drawdown, they did dampen the enthusiasm of state and federal agencies. Legend's agitation continued well into 2005. In addition, he rallied support from some private environmental organizations to continuously hinder progressive projects, such as construction of a new and more efficient spillway and dredging operations that would clean out clogged boat channels. He prided himself on knowing laws and regulations instead of knowing the ecology and needs of the lake.

As of this writing, Legend's lawsuits have yet to be settled. The riparian rights suit has called public ownership of the lake into question—a potentially devastating threat that would leave only four thousand acres to the state, according to some elected officials.[1] The duck blinds lawsuit, meanwhile, was the culmination of a contentious issue that had probably caused more strife at Reelfoot Lake than land ownership, fish regulations, and lake levels combined. It arose in the 1980s after hunters were allowed one year to claim and register a chosen hunting site, after which those sites would be declared permanent and no new sites could be claimed. For a hundred years prior to this, hunters had been claiming and reclaiming duck blinds and duck-hunting sites, and private wars, some involving felonies, often erupted over these sites. It had gotten to the point where the use and management of the lake all seemed to revolve around the use and location of the blinds, even though they were seldom used for more than two months out of the year. We managed to burn and destroy more than a hundred ramshackle blind structures, but how we did it without causing a general uprising is a mystery to me. Our registration plan was supposed to forever settle the issue, but the squabbles continued and became connected with the riparian rights issue.

The Tanner-Hamilton Team

By the mid-1980s, "Mr. Legend" appeared to be the most forceful representative of issues of local interest. No one else had come forth with a similar energy and dedication. Meanwhile, not since 1925 had political opportunities for Reelfoot Lake been better. Anyone having significant knowledge of the lake knew that time to save its natural qualities was running out. "This is the last opportunity that mankind will have to preserve and maintain this lake," warned State Representative John Tanner in 1983, "because it is filling now so rapidly that any more bureaucratic delays will result in any action being too late."[2]

Tanner and State Senator Milton Hamilton, both Democrats, were the perfect candidates to help Reelfoot. Both legislators were from Obion County, where more than half of the lake was located. Both had property interests at the lake, and both understood its value: one survey would show that 212,645 visitors spent over $25 million annually on trips to Reelfoot.[3]

Senator Hamilton, after Governor Ned McWherter, was arguably the most powerful politician in the state. He held sway over the appropriations committee, and at the time the state had a healthy reserve. Hamilton was anxious to divert whatever was needed and available from the state treasury for the betterment of Reelfoot Lake. Representative Tanner had another motive: he was running for a seat in Congress and needed a high-profile issue to champion at this stage of his political career. He thought he knew the hearts and minds of the people and could persuade them to accept the drawdown for the greater good of the lake.

Together, Tanner and Hamilton were a powerful team during this period, with the ability to route millions of dollars to the lake and its people. A vote of confidence from the public, even a narrow majority, would accomplish a great deal for this neglected part of the state and create momentum for a new direction in wetland management for a long time to come. The timing was right, the determination was right, the reasons were right, and the money was right. All these state legislators and the TWRA needed was a symbolic local leader to stand up and request—no, demand—that certain things be done to improve Reelfoot Lake. Legend appeared to be that person, and his start was impressive.

Through his own initiative, Legend had special influence on a 1983 legislative task force for Reelfoot, and in no time, he had gained political support within and outside the area. Our own administrators cultivated his support and listened to his recommendations and concerns. He also helped hurry several bills through the state legislature that appropriated funds for TWRA projects at the lake. Eventually, he was recruited to take a departmental position with the Department of Environment and Conservation. His energy appeared unstoppable, and he and I had numerous late-evening discussions about the lake's needs. No doubt, he had great compassion for the lake and appeared to be genuinely committed to improving it. He attended meetings with Gary Myers and Harold Hurst and was on the inner loop of Senator Hamilton's or Representative Tanner's planning sessions and late-night brainstorming. Thousands of dollars were appropriated. Some money was spent questionably on Legend's advice alone, but many projects were obviously beneficial.

Nevertheless, the riparian rights issue—a diversion that deflected attention from the ecological needs of the lake—was the one most important to Legend. Before long the issue dominated most of our discussions, and it became a major stumbling block to the support needed to implement the drawdown. At one point, exasperated, I gave him some advice: "Whatever is Caesar's is Caesar's; whatever belongs the state is the state's. The riparian issue is yours and the courts'; TWRA's goal is to improve the biology and wildlife recreation potential of Reelfoot Lake." My concern touched a sensitive nerve, but it lasted only a short while. I offered no more comment on the ecological issues, and the topic was rarely mentioned again—that is, until I was later subpoenaed for testimony in one of Legend's lawsuits.

Unfortunately, Legend continued to aggravate Paul Brown or me about the day-to-day management of the lake. Soon he had alienated Paul, and communication between them nearly came to a halt. His effectiveness for our cause soon vanished. He fell from the grace of our agency, but not as quickly as he had gained it. His tenure with TDEC was also short-lived. Once again, we were without effective local leadership.

In the end, we had consent from key officials needed to accomplish the drawdown, but we did not have the blessings of the local people. All of the public meetings, visual aids, videos, and reams of research and supporting documents did not seem to matter in the end. Privately, many locals encouraged us to continue our work to improve the lake; publicly, though, they kept their views quiet. Ultimately, a rough consensus of the vociferous and misguided few—whatever their good intentions—had the final say about Reelfoot's future.

The Drawdown Begins

On May 26, 1985, the Reelfoot Lake drawdown was one day shy of its scheduled start. A year had elapsed since Director Myers authorized the plan. Meanwhile, local opposition against it had brewed to an untenable level. The schedule called for opening the spillway gates at 8 a.m. But to insure that the disturbance had not reached the level of a political crisis, Paul and I had a conference call with Myers and Region 1 manager Hurst.

Myers was still with us, but the dark cloud of controversy loomed even heavier within hours. The press showed up with a renewed interest; television reporters clamored for the latest news. A small local crowd gathered at the spillway with a single purpose: they wanted headlines. Paul and I had already received inquiries from the media, but our answers failed to satisfy them. Explanations based on our technical knowledge were much too complicated to reproduce in thirty-second sound bites. It was the perception of civil unrest that made news. It appeared that the government was about to force its will on the people at Reelfoot.

The procession of objectors at the spillway numbered about a dozen at the time, and some toted a morbid mascot: a black casket with black roses. Bearing the words "The Death of Reelfoot Lake," the casket was paraded back and forth across the spillway. Meanwhile, a small crowd gathered, and the activity continued into the evening.

Where, I wondered, did these folks get the notion that the lake would die? Did they really think that we intended to destroy it? And on whose expertise had they come to this conclusion? These questions were not up for consideration. Most had already made up their minds: "draining" the lake would ruin it, and it had to be stopped. The crowd grew. I believe that the entire gathering could have been dispersed if someone had announced that a mud-wrestling match was underway just down the road. But I became concerned that we had not adequately described the growing unrest and called Myers a second time. The director did not waver: "It is the right thing to do for the lake and the citizens of the state. We ought to do it."

After dark, word got out that a group of gray-haired ladies intended to gather below the spillway the next morning. Of course, anything or anybody would be washed away downstream in the avalanche of water that would be released. The news media had a keen sense about these matters. The threat had to be taken seriously, since embarrassment to the TWRA or injury to the protestors could occur. Thus, we reconvened again with Myers and Hurst and suggested a revised schedule. Instead of waiting until daylight, Paul and I decided to initiate the drawdown at 2 a.m. Myers and Hurst agreed. That way, the drawdown would at least occur on the day it was announced, albeit at an early hour.

Wendell Crews, manager of the Reelfoot National Refuge, was briefed on our plan. He agreed that the state should assume total responsibility for the drawdown and cooperated by giving up the crank necessary to operate the spillway gates. (Handing over the crank to TWRA proved to be a key factor in later legal action.) The late-night detail included Paul Brown, three or four local wildlife officers, and me. Professor Wintfred Smith of the University of Tennessee at Martin was asked to join us. He had credentials as the lake historian, and he was also on the UTM research team that investigated the biology of the lake.

Armed and in field uniform, we looked like a SWAT team when we showed up at the spillway for the 2 a.m. gate opening. There were reasons for our quasi-military appearance. A recent shotgun blast had blown out windowpanes at the WMA office. In addition, there had been ominous threats from locals mentioning scoped rifles and dynamite.

The gates, however, were opened without incident, and our mission was accomplished within ten or fifteen minutes. There were no shots except those from Smith's camera. Not another soul was there to witness the early morning event. Indeed, the drawdown was historic. Nothing this dramatic had been done to affect the lake since its creation by the New Madrid Earthquake. What we could not predict was that the ensuing public discontent would, arguably, be even more dramatic and extend over the next couple of decades.

The Lawsuit

About a week after the drawdown began, a lawsuit challenging it was filed in the U.S. Federal Court in Memphis. Judge Odell Horton would hear the suit. Several TWRA and USFWS personnel were subpoenaed for a June 24, 1985, hearing. Wendell Crews, Paul Brown, and I were among them.

The litigants were business owners and local politicians living around the lake. The lawsuit was heard on the grounds that the drawdown proposal was a major federal action, as defined by the National Environmental Policy Act, which also changed the course of events in the West Tennessee Tributaries Project. Ironically, as in the WTTP suit, a law enacted to protect the environment was being used to for the opposite purpose. The lawsuit had nothing to do with improving Reelfoot Lake. In my opinion, it was a legal effort by the litigants to defy government "interference" with a lake that they felt belonged to them alone. The plaintiff's lawyer knew that if he could involve the NEPA, he might stall or permanently eliminate the drawdown; thus victory could be declared for his clients. The outcome could decide the fate of the drawdown, if not of the lake itself. The lawsuit's central question was whether the U.S. Fish and Wildlife Service had any role in the drawdown. If so, the next question was whether the project would have a major impact on the environment. If the answer in each instance was yes, this would trigger the NEPA, requiring the Fish and Wildlife Service to complete an environmental impact statement. If the court found that the federal government was not involved, then the state could proceed with the project.

The state's position was that the drawdown was a state project and did not involve the Fish and Wildlife Service. It held that Reelfoot was the state's property and that the state—not the public or the USFWS—had the primary right and obligation to manage it. The decision was up to the federal court judge.

Being a defendant in federal court can be unpleasant. It certainly was for me when I was cross-examined by the plaintiffs' attorney, Alf Adams. My testimony lasted from 10:15 a.m. to 3:30 p.m. on the first day of the trial. TWRA's attorneys, Mary Walker and Mike Perigren from the State Attorney General's Office, were excellent lawyers and nurtured the case as if it were their only child. Being unfamiliar with legal procedure, I saw Adams's strategies as offensive. The cross-examination seemed to downplay our intentions and hard work. Despite encouragement from Mary Walker, I did not excel as a witness. Reelfoot had become too personal for me to remain stoic about our hard work, and Adams's relentless questioning caused me to answer with indignant impatience. Judge Horton took a dim view of my posture.

Wendell Crews also took the stand. In contrast to me, he went through the process with the ease of an ancient sage, which in a way he was. He had seen it all before. Wendell became our best hope for getting through this legal sparring. Even in the face of Adams's badgering questions, Wendell's humility and honesty were unshakable, and he gained the admiration of the judge.

Unfortunately for us, a week later, the message came from Roy Anderson, TWRA's chief of wildlife, that Judge Horton's decision entirely favored the plaintiffs. The presence of the U.S. Fish and Wildlife Service and the act of Wendell Crews in handing over the operation of the spillway gates were sufficient to prove federal involvement. In addition, the drawdown was declared to have a major impact on the lake. Thus, the NEPA requirement for an environmental impact statement was activated.

An injunction ordered the spillway gates to be closed; the drawdown was finished. A rare opportunity for the people at Reelfoot Lake had been missed. Since the lawsuit was a federal action, the task for preparing the EIS fell to the U.S. Fish and Wildlife Service, and a grinding, expensive process soon began.

Whether it was from apathy, or some reluctance to become fully informed, or from poor representation allegedly on their behalf, the people depending on the future of the lake did not grasp the missed opportunity. We asked our experienced Florida biologist friends what to do next. "Forget it and go to a lake where people will support you," they said. But we had only one Reelfoot Lake. That was reason enough to try a little longer to save it.

Chapter 14

Wrestling with the Fate of Reelfoot Lake

So, following the court decision, it was back to the drawing board. At the suggestion of Gary Myers, Paul Brown and I did further research on lake drawdowns, hoping to avoid the failure of any future effort. We made field trips to Florida, Louisiana, and Arkansas in the fall of 1985 and the summer of 1986. In Louisiana, many of the lakes we visited were similar to Reelfoot: they had cypress trees standing mid-lake, common lotus, duckweed, and the like. In Arkansas, we saw how biologists systematically used drawdowns to control noxious aquatic vegetation and thus improve the fisheries. By rotating the schedule, the Arkansas fish and wildlife agency used this technique on two dozen lakes each year as a matter of routine practice. All of these trips were beneficial, reinforcing the credibility of the Reelfoot drawdown plan.

If the result of the drawdown lawsuit was a grave error for the future of Reelfoot Lake and the people who used it, it was also a test of the partnership between the TWRA and the U.S. Fish and Wildlife Service. The long, dreary process of jointly developing an EIS was not finished until 1989 and cost, according to estimates, more than a million dollars. Since the courts, citing the National Environmental Policy Act, considered it a federal project, the USFWS was responsible for completing the EIS. In reality, it was a state project, and the USFWS was a disadvantaged and reluctant partner.

Reluctant or not, the Fish and Wildlife Service pulled its share of the load and became a strong proponent for the restoration of Reelfoot Lake. In May 1986, for example, a group representing both the TWRA and the USFWS's agents in Tennessee visited the USFWS's Western Energy and Land Use Team (WELUT) at their database center in Ft. Collins, Colorado, to go over many of the problems we encountered during the previous year's aborted drawdown attempt. Following this meeting, WELUT scheduled workshops to address alternatives for Reelfoot. These were held August 24–29 at the Reelfoot Airpark Inn and continued in Memphis. From these workshops the WELUT team provided significant help in formulating the EIS and our other plans for the lake. Yet, even with such cooperation between TWRA and the UFWS, the process proved taxing to both agencies.

A Burden for the USFWS

The USFWS's Atlanta office was certainly not pleased with the State of Tennessee for pulling them into this legal quagmire. The fallout went deeper than the taxpayers' pockets, public opinion, or the lake itself. At no time was the relationship between the USFWS and the TWRA more tested than during the period from 1984 to 2002. Prior to this time, the national wildlife refuges had more or less been isolated "federal islands" coexisting with the state lands; the federal agency alone managed wildlife on its own land. But this was not the best way to manage migrant birds or any other wildlife, since wildlife are not constrained by survey boundaries. While the national wildlife refuge had jurisdiction over migratory birds, Tennessee had jurisdiction over resident wildlife. The federal agency's natural tendency to isolate itself from state involvement made for a cordial but limited relationship between USFWS and TWRA for years.

The EIS required by Judge Odell Horton forced us to face each other at the close quarters of a conference table. Every conflict our agencies had ever known probably surfaced during the joint EIS project. Yet, it was an important litmus test for two wildlife conservation agencies dedicated to a common cause.

The USFWS personnel assigned to the EIS would probably have preferred a daily beating to having to complete the assignment. We felt the same way, since many of the drudgeries would befall the TWRA staff. Nevertheless, the USFWS had to accept the brunt of Judge Horton's decision. The conditions outlined in the 1941 lease agreement also required it, and the lease would not expire until 2016.

The lease agreement had put the USFWS in charge of managing water levels at the lake, a responsibility not easily affected by state or local political whims, and for the most part this arrangement had been favorable both to the state and to federal refuge system objectives for managing migratory birds, especially waterfowl. According to Judge Horton's ruling, the USFWS had abrogated this assignment to TWRA with the drawdown on May 27, 1985. But in our interpretation of the lease agreement, water-level management of the lake also required any change to be in concert with the state's management. Thus, when the Fish and Wildlife Service allowed the TWRA to operate the spillway gates for the drawdown, it was with the hope of both agencies that the drawdown would be seen as an action by the state alone. The reasons for initiating the drawdown were ours, not those of the Fish and Wildlife Service, since they were otherwise satisfied with their programs.

The plan for manipulating water levels of Reelfoot Lake in the drawdown plan was in fact a disruption of the USWFS's normal program for managing waterfowl, albeit one they agreed to in a spirit of cooperation with our agency. With the result of the lawsuit—Judge Horton's decision that the USFWS had improperly acquiesced to state objectives—the partnership between the USFWS and TWRA was welded together. Now we were stalled in the same bullpen and had to suffer the consequences simultaneously.

At first, the USFWS's Atlanta office was reluctantly amenable to making concessions in their management programs to accommodate the drawdown, but the lawsuit caused more encroachments on their operations than could be tolerated. An unavoidable discord developed because of it. If the EIS was not enough to tax the USFWS personnel, they were even more discomfited by pressure from "Mr. Legend," who—even before the drawdown controversy developed—had charged the agency with not doing its duties set forth in the 1941 lease agreement; he criticized the way it managed water levels, cleaned out boat trails, and provided funds and assistance for various other lake projects. His persistence instigated a movement through the state legislature to change the lease. TWRA was sometimes inclined to agree, since the agency's Atlanta office had been reluctant to cooperate on several occasions.

In October 1983 this agitation had resulted in an effort to rewrite the 1941 lease agreement. In fact, a revision was drafted and discussed at some length by both parties, which led to an unfortunate but temporary rift between Harold Benson, USFWS's assistant regional director, and the TWRA. The USFWS was not accustomed to giving up anything, and tinkering with the lease agreement was enough to spark Benson's anger. His Atlanta office considered abandoning USFWS responsibilities at the Reelfoot refuge altogether, but this thought was short-lived.

Once our differences were patched over, we began work on the problems at hand. In fact, the USFWS did have a significant stake in the lake and how the EIS would be concluded. On the one hand, the state owned the entire twenty-four thousand acres of public land in and around the lake within the Tennessee borders, which included nearly all of the open lake. On the other hand, the federal government leased some eight thousand acres from the state as a refuge for migratory birds, except at the site around its headquarters near the Walnut Log community, which was on land they owned outright. Some two thousand acres, mostly marsh and lowlands, were also owned by the federal refuge at the north end of the lake in Kentucky. Under these circumstances, it was not difficult to think that the State of Tennessee had superior rights to the water levels of Reelfoot Lake, albeit in consultation with the USFWS and Kentucky. But according to Judge Horton's decision, the lease agreement had legally placed the USFWS in charge of the spillway gates at Reelfoot Lake and thus the control of water levels.

To circumvent the legal obstructions, the state went so far as to consider removing the obstacles that had originally triggered the NEPA process: TWRA proposed a land swap to move the Reelfoot National Wildlife Refuge out of the waters of Reelfoot Lake, thereby removing federal jurisdiction from Reelfoot and making the NEPA inapplicable. "How about building a levee just south of the national refuge boundary?" I suggested. "It would separate the refuge and the Kentucky farmers from the waters south, or 90 percent of the lake TWRA wanted to manage."

This idea did not work, as such a decision would require its own EIS. There seemed to be no end to the legal quagmire into which we had stepped. We had to either get along or wait until 2016, when the lease expired, and then start a new arrangement.

But the trying times passed. In 1989 the USFWS finally completed the EIS with TWRA's participation. The EIS not only recommended the Reelfoot drawdown but strengthened and expanded the project. In the end, the difficulties led to a stronger alliance between the state and federal agencies, and it eventually spread to include all of the federal refuges in West Tennessee. This new cooperative spirit resulted in numerous joint projects that, otherwise, would likely never have happened.

In 1987 Wendell Crews retired from the U.S. Fish and Wildlife Service, and Don Temple took his place. Temple was a bold and practical manager unafraid of taking risks. His steadfastness solved numerous sticky and aggravating problems. In December 1989 Temple transferred to another refuge, and Randy Cook became the new refuge manager.[1] From our perspective, we were fortunate to have the cooperative companionship of these competent and professional managers.

The Reelfoot Lake Fifty-Year Plan and the EIS

As I have suggested earlier, "Mr. Legend" was almost a one-man initiative. Among his accomplishments, he had spurred the 1986 General Assembly to pass Joint Senate Resolution 235, which required TWRA to produce what became known as the "Reelfoot Lake Natural Area and Fifty-year Management Plan." The plan, of course, was flexible and designed for practical implementation and to establish the history and management philosophy for Reelfoot Lake. But Legend advanced the plan to fit his own theories for how the lake should be managed. For example, he wanted the lake managed as a "natural area" in its strictest sense, which would have put it under jurisdiction of the Tennessee Department of Environment and Conservation and restricted its use as a WMA. However, TDEC chose not to exercise this option—a wise decision, in my opinion.

Planning fifty years ahead was an unprecedented but smart thing to do. To say that Reelfoot Lake needed a long-range plan was an understatement. But the lack of an organized local voice at Reelfoot and the lack of interest in long-range planning were reasons to be skeptical about putting too much effort into something that looked fifty years ahead. Any such plan would collapse without at least two things: (1) specific leadership (that is, a person charged with seeing that the provisions of the plan were followed); and (2) a formal committee broadly representing those affected—a group that would agree on major changes and ensure that the leadership for implementing the plan was held accountable.

For the next four years, Paul and I wrestled with input and revisions, and the contentious issues involving the 1941 lease agreement. Finally, in 1988, the "Reelfoot Lake Fifty Year Management Plan" was competed, just one year before the USFWS's environmental impact statement on water-level management for the lake would be finished. This 310-page report, with its broadly based goals and objectives, sought to "preserve and enhance traditional outdoor recreation opportunities at Reelfoot Lake and preserve Reelfoot Lake for its intrinsic values as a unique ecosystem."[2] It was, in effect, an umbrella plan more concerned with long-range goals of various programs (including water-level management, wildlife and fisheries management, and aquatic weed control) than with their specific details.

The EIS, of course, did address specifics, at least insofar as the proposed drawdown was concerned.[3] Acceptance and implementation of the EIS meant that the lake would be lowered four to eight feet as dictated by environmental indexes. It was estimated that the resulting improvements to the lake would significantly increase the annual income from recreational facilities and use of the lake—an increase of $2.5 million for a four-foot drawdown and $2.8 million for an eight-foot drawdown. (This latter option would require an improved spillway since the existing one could not lower the lake much more than five and a half feet.) The annual value for full implementation of the EIS was expected to exceed $13.5 million.

The EIS contained other good features that promised to enhance the lake's health and benefits. A thousand-acre sediment-retention reservoir was proposed for the lower end of Reelfoot Creek near the national refuge boundary. This reservoir would capture 70 percent of the sediments entering Reelfoot Lake and double as a waterfowl area. A new spillway was proposed to lower and raise the lake to the extent needed; it would also be designed in such a way as to boost the natural stocking of the lake with fish. Also, in-lake channels were designed for water circulation—a feature that would not only improve boat access but also the movement of migrating fish throughout the lake.

Shelby Lake (a part of the area known as the "Scatters"), a historical fish brooding and waterfowl wetland below the spillway, had been drained and farmed. A thousand-acre wetland was proposed at this location to restore part of the former wetland. This wetland would then act as a fish-breeding and -holding area where a timely rise in the Mississippi could allow fish from the hatchery to enter the lake. In addition, the developed wetland would provide a migratory bird management unit.

The EIS satisfied NEPA requirements and quieted the critics. At the very least, it set in motion an interim plan that has provided for more dynamic water levels at Reelfoot Lake since the early 1990s. This part of the EIS recommendations was approved by the Memphis district federal court and cost the public very little. This plan involves the implementation of a temporary water-level management technique that allows the lake to fluctuate higher than the stable—and detrimental—artificial water levels of the past. Under this plan, water levels are allowed to fluctuate one foot above the normal pool level from November 15 through April 15 and one-half foot higher for the remainder of the year before the spillway gates are opened to release flood waters. Doubtless, this management has greatly improved waterfowl hunting and fishing over the past two decades. At least five thousand acres of fish and waterfowl habitat is temporarily added to the lake during the implementation of this plan.

Opposition from County Governments and Landowners

After the EIS was completed, activity of all kinds picked up within the USFWS, the state legislature, and the TWRA. One initiative was to purchase or reclaim land necessary to implement the drawdown, because the EIS water-level management plan called for higher water in the lake following the drawdown. TWRA director Gary Myers, with the help of others, managed to get a loan from the state general fund to start the purchase of a four-thousand-acre buffer zone around the lake that would allow the water to be raised 1.8 feet above normal. We believed that this elevation would cause minimal impact to private lands, although Kentucky farmers disagreed: they argued that as many as four thousand acres of farmland would be eliminated from production. (According to press reports, however, later scientific studies sponsored by the Corps of Engineers would show that even in a worst-case scenario involving heavy rainfall, no more than fourteen acres of farmland would be flooded—not that this made much difference to the Kentuckians.[4])

More than twenty-four hundred acres from thirteen tracts were purchased by the state in the buffer zone by July 1991. It did not please the county governments in Lake County, Tennessee, and Fulton County, Kentucky, or the landowners there. They envisioned no redeeming benefits to the lake or their counties. Even though Lake County was eventually compensated for lost taxes, it was not enough. The subject continued to be a topic of complaint on the county commission's agenda. Ironically, thirteen houses were purchased and razed at Walnut Log, Tennessee, to accommodate the high-water expectations at the lake. A way of life went with the houses, but those folks complained the least.

A small group of Kentucky landowners with agricultural interests opposed any purchase of land, whether it was within a flood zone or not. A lucrative price was paid to willing sellers with farmland, yet even these people complained that the government was taking their land. They despised the wildlife agencies and anything associated with the drawdown plan, although hardly any private land in Kentucky would be flooded. Indeed, the Kentucky landowners opposed any management of Reelfoot Lake, except to subdue it to their liking, and they lobbied through the Farm Bureau, a well-known insurance company, to halt the drawdown proposal. The State of Tennessee had little means to purchase land outside its boundaries, and the Fish and Wildlife Service's Atlanta regional office lacked the initiative to do so, probably because of the ever-present political pressures.

The Kentucky Department of Conservation was enthusiastic about our drawdown plan in the early 1990s. It attempted to purchase a shallow lake near Reelfoot in order to supplement their conservation objective, which was to restore and enhance wetland habitat used by migratory birds and other wetland species around the region. This lake was only a short distance north of Reelfoot, and the purchase would be from willing sellers. This plan promised to complement the overall management of Reelfoot Lake and benefit Kentucky sportsmen and other conservation-minded citizens, as well as help solve the flooding concerns. However, according to our Kentucky counterparts, when word reached the local politicians in Kentucky, the Kentucky Department of Conservation was suddenly instructed by its governor's office to withdraw from the Reelfoot projects and stay out of Fulton County. It did so. Its representatives' regular attendance at the Reelfoot Technical Committee meetings ceased, and they have not been heard from since.

The Corps of Engineers' Plan

Meanwhile, the water-level management plan proposed in the USFWS's environmental impact statement faded to a dim memory. The Fish and Wildlife Service had satisfied all of the legal requirements, addressed the environmental questions, and given alternatives. They had tried to clear the air about everything that would move the management needed for Reelfoot Lake forward. But around 1993, when the multimillion-dollar project was ready for implementation, there was no money available (some say because of Kentucky politics) and little hope that the state could generate it. This and rising objections from the Kentucky landowners encouraged the U.S. Army Corps of Engineers to become involved.[5] The corps was more than willing to come to the rescue, since by this time the WTTP had been shut down and it needed work to do. But, as always happens whenever the corps becomes involved in a project, it would have a strong say, if not the final word, about how that project would be managed.

A role in the management of Reelfoot Lake placed the corps in the precarious position of being peacemakers in the risky business of compromise, and such an approach often blunts the original objective. The county levee boards constitute one of the corps's most powerful political arms along the Mississippi, but the boards in Lake County, Tennessee, and Fulton County, Kentucky, were strongly tied to the county fathers and the farming industry and neither had much interest in Reelfoot Lake. This lack of concern is understandable to some extent: after all, the farmers' and perhaps the county fathers' livelihood depended heavily on crop production, not on natural resources such as fish and wildlife, wetlands, and trees. So, it was no great surprise that the farmers in Kentucky tended to be skeptical about the corps's partnership with Tennessee on the Reelfoot restoration project.

Nevertheless, the corps bravely entered this arena of controversy, and the central feature of its involvement at Reelfoot would be the construction of a new spillway at an estimated cost of $35,287,000. Complaints from Kentucky farmers about the Fish and Wildlife Service's plan led the corps ultimately to decide that a new, full-blown EIS was needed—though whether a new statement truly was needed is highly questionable. In any event, three different reports would come out of the process: a reconnaissance report, a feasibility report, and, finally, another environmental impact statement. All of these would consume time and money, but the EIS was by far the most complex, requiring several years to complete. Common sense should have made it clear that these expenditures of time and money were unnecessary even before they started.

As the controversy dragged on, the corps and the State of Tennessee proceeded with the partnership. The reconnaissance report was completed in 1993, followed by the feasibility report in 1995. The final EIS would not be released until December 1999.[6] Just a few months before that, authorization to proceed with construction of the spillway had come when President Bill Clinton signed the Water Resources Development Act of 1999, but various disagreements continued to dog the project.

From the perspective of the TWRA and the USFWS, the extent of the corps's authority over management of the lake was a major bone of contention. The corps, ever focused on flood control, wanted total say over how the lake levels should be managed, and their plan—designed to accommodate the complaints of the Kentucky farmers and the politicians who represented them—was disastrously inadequate, in the opinion of many of us. This all became clear at a series of meetings during the summer of 2002. The first meeting, which included Gary Myers and Randy Cook among the attendees, took place in late

June at the Washington office of John Tanner, now a congressman, and its purpose was to reach agreement on "language" that would allow the corps to proceed with the spillway construction. Speaking for the USFWS, Cook strongly objected to any agreement that would allow the corps to control its operations and the water levels at Reelfoot. Tanner, however, emphasized the need for the spillway to be built, saying that the disagreements over its operation could be worked out once the construction was complete. Myers reluctantly agreed, and that seemed to settle the issue for the moment. Cook called me shortly afterward and said, "Reelfoot Lake died last Wednesday."[7]

A second meeting, which I attended, was held on July 18 at the Reelfoot WMA. I pointedly asked Colonel Jack Scherer, the corps's district engineer, what the corps's plans for water levels would be during an extreme lake drawdown. He said that the drawdown would be three feet, perhaps four, if all went well. But this, I responded, would not address the lake's biological ills. I pointed out (1) that the extent of the drying would be insufficient and (2) that the water would not rise high enough upon refill to cover and kill the vegetation, such as willows and giant cutgrass, that was expected to grow during the drying process. When Scherer asked what we needed, I replied that the lake should be lowered at least six feet and then raised a foot higher than normal pool on refill. Randy Cook immediately supported this view.

The corps staff was dumbfounded, knowing that the Kentucky landowners would have none of this—even though the corps was already aware that the USFWS's 1989 environmental impact statement had proposed this drawdown schedule. Another attendee at the meeting—Greg Wathen, Myers's assistant—was taking notes and, to his credit, announced that he would tell the TWRA commission that Cook and I did not agree with giving up the state's rights to manage Reelfoot Lake and, also, that we recommended dropping the corps's participation in the project. Our contention was that since Reelfoot was not considered federally navigable, the corps had no right to regulate its waters—unless we signed an agreement to give up those rights. And that, we said, was something we could probably not do. That ended the meeting.

At a third meeting, held at the corps's district office in Memphis, Cook and I repeated our arguments. Myers, however, was in attendance this time, and he said that we ought to agree with the modifications proposed by the corps and the landowners. He still agreed—not unreservedly, I am sure—with Congressman Tanner's argument that the important thing was to get the spillway built before trying to resolve the question of who should control the water levels at the lake. And indeed, the proposed spillway was one good thing in the corps's EIS. The corps recommended an automated, state-of-the art design that had the capability to assure that all water-level management objectives could be met. Lake levels could be drawn down on demand to targeted elevations; as much as eight feet of water could be drawn down within forty-five days. The new design would also allow fish from the Mississippi to enter the lake more often than the old spillway had done. This aspect of the design had the potential to greatly rejuvenate the Reelfoot fishery by providing forage fish and replenishing the native species that were once characteristic of the lake.

But to put off the management issue in order to get the spillway built was just asking for trouble later on, in my view. The objections of the landowners—their fear, however unfounded, of losing valuable farmland—would be the same. When (or if) anyone again mustered the energy and will to tackle management issues at Reelfoot Lake, another EIS would undoubtedly be required. And heaven only knew where the money for that would come from and who would undertake it. Randy Cook and I pondered the results of these meetings for several months. Still unsettled about the situation, we sent a note to Myers early in 2003, listing the reasons we believed he should renege on his agreement to go along with the corps's plan.[8] If our concerns mattered, we never knew.

Some months later, however, the state did opt to exclude the corps in constructing the spillway. Hoping to skirt the corps and the demands of the Kentucky farmers, the Tennessee Department of Transportation was recruited by John Tanner to replace the old spillway on Highway 21 and build the newly designed spillway nearby. The package was eventually approved and funded, and the intended construction was announced in early 2004. Infuriated by this move, the Kentucky farmers began a new campaign with the support of U.S. Senator Mitch McConnell. For his part, Representative Tanner was infuriated by Senator McConnell's audacity in trying to stop a project he had diligently worked on for years. Tennessee State Representative Phillip Pinion and State Senator Roy Herron gathered forces with Tanner for a duel with McConnell, and that duel still continues as of this writing.

The Progress of the Plans, 2005

The extreme drawdown of Reelfoot Lake that was attempted—and aborted—in 1985 became a rallying pole for all the ills and frustrations that had accumulated in the area since the 1925 Reelfoot Commission attempted to secure the preservation of Reelfoot Lake for Tennessee's citizens and visitors. Hundreds of studies, environmental reports, pieces of legislation, lawsuits, and public meetings—not to mention millions of dollars and innumerable man-hours by individuals and state and federal agencies—have resulted in few material improvements for the management of Reelfoot Lake. In fact, nothing on the immediate horizon indicates major improvements for the ecosystem of Reelfoot, save one thing: the proposed new spillway that has the capacity to accommodate future water-level management programs. Unfortunately, the fate of this badly needed project is at best tenuous—held hostage by politicians and Kentucky farmers for the sake of less than a hundred acres of marginal farm ground. As far as these people are concerned, the only good water-level management is that which is most favorable to them. Meanwhile, the preferred management plan, scrutinized and sanctified by the courts, has been collecting dust.

Adding to the uncertainty have been the still-unresolved lawsuits over riparian rights and duck blinds at Reelfoot. As Representative Pinion and Senator Herron told reporter John Brannon of the *Union City Daily Messenger*, these suits have threatened the very notion of Reelfoot as a public lake. "I am also very concerned," said Pinion, "what effect [an adverse ruling] would have on a new spillway, especially since Congressman John Tanner, Gov. Phil Bredesen, state Sen. Roy Herron and myself have worked so hard to get funds to build it."[9]

Despite all the ups and downs in the struggle for Reelfoot, there has been one redeeming, if temporary, spin-off: the interim water-level management plan that I mentioned earlier. Unfortunately, very few fishermen and waterfowl hunters have understood the importance of this technique, or why they have enjoyed more successful outings since the plan was implemented. In fact, now that it has been in effect for more than a decade, it may have caused them to become complacent. If so, they should keep up their guard. The fact that fishermen now catch good-sized fish in good quantities and that hunters can gain better access to their duck blinds is not a guarantee that all is well. They should continue as a unified voice to aggressively support active management for Reelfoot Lake, whether it involves a drawdown or any other help from lake managers to improve their outdoor recreational activities.

Kentucky farmers and landowners are not likely to change their minds, since they have rejected government offers of compensation for perceived damages that could be caused by the management of the lake. It has mattered not that their choices have diminished millions of use-days for hunters, anglers, and others who enjoy the lake. Their business is farming, and in their minds, this includes the Reelfoot floodplains.

For wetland managers, the experience at Reelfoot Lake provided valuable insight—ecological, sociological, and political—into wetland issues. Thus, as we tackled parallel issues with our projects along the Obion and Forked Deer rivers, what we encountered came as no surprise.

Chapter 15

Reviving the Dwindling

Interest at Tigrett

By the end of January 1991, interest in Tigrett WMA had slipped to a new low from weariness. The first draft of the restoration plan had been under scrutiny for about seven years, and the area it affected had been reduced to about a quarter of what the original plan had proposed. There was little left to talk about by anyone but the Stokes Creek farmers. So, it was no surprise that after a poor outcome from the review committee, Gary Myers's release of the revised "Tigrett WMA Floodplain Restoration Plan" raised no eyebrows.

Even the Tigrett Technical Committee lost interest, and fewer meetings were held. The entire subject of river restoration seemed to be at its end. One or two on the technical committee joined the Tennessee Conservation League in condemning the plan. "It lacked scrutiny from civil engineers," some said. As usual, there was no support saying, "The idea of river restoration is good—let's make it work." But there was no shortage of critics finding ways it would not work. So far, nothing had been done to dry up one artificial swamp, cause one stream to start flowing, or provide for one living tree, duck, deer, rabbit, or shorebird.

Officials at the Tennessee Department of Environment and Conservation decided that the Tigrett project contradicted the nationwide policy for wetlands. They reportedly considered rewriting water-quality regulations to head off any attempt to drain artificial swamps in river floodplains. Frustrated, I asked Dan Sherry, "What sense does it make to despise the WTTP, loathe channelization, and yet support the swamps this very activity helped create?" But Dan was unsure. "Maybe they are right, maybe they are wrong, I don't know," he said. "The conservation community already has questions about your allegiance to the nationwide movement to save wetlands. It's your position on the artificial swamps. They see it as destroying wetlands." Whatever was going on in their minds was certainly not from listening to our arguments; they never got that far. They had certainly misjudged our intentions. Salvaging wetlands was something we intended to stake our careers on. I began to wonder how it could be that we were so much out of sync with these people. I think we were still considered "loose canons"—overzealous land managers, remote from the real world. The problem, as I saw it, was that the political process rested contentedly on its ignorance. The makers of law and policy had no idea how important their role was for the future of natural resources in this country. This time, they had neglected some of their most sensitive constituents—hunters and fishermen. Next time, they might neglect an even bigger group—registered voters in the general populace.

The Tennessee Conservation League—which our agency saw as the primary, organized voice of the state's sportsmen—supported the Tigrett project at first but then reversed its opinion. Tony Campbell, the TCL executive director, squirmed when I tried to argue that his group, of all people, should support the Tigrett plan. One of his major objections was the draining of the artificial swamps. Regarding the Tigrett plan, Campbell wrote in a letter to TWRA field director Ron Fox: "We expect the drainage of the large areas of ponded shrub marsh to result in an overall reduction in WMA species richness and biological productivity to the detriment of waterfowl and other wetland dependant species."[1] But the reverse was true. We had either failed to communicate, or he had another reason for saying this.

Campbell apparently had not bothered to consider the investigations completed by our fishery biologists, or that done by Professor Neil Miller from Memphis State University. Their studies had proven his point about the "shrub marsh" to be wrong. Campbell seemed only to support duck hunters who liked these swamps, no matter how bad they were for the ecosystem, his constituents, the local landowners, or other hunters. This perception—that artificial swamps were good for waterfowl—served either ignorance or another purpose. For any capable wildlife biologist, it suggested something not kosher on the other side. "There's been a lot going on you don't know about," Campbell told me at an October 1990 meeting.

Indeed, there was a lot going on that I did not entirely understand. In fact, a decision had already been made among J. Clark Akers and his circle of advisors that they should not waver from their insistence that the WTTP mitigation land be developed as originally proposed—which meant more levees, pumps, and water control structures, much like the traditional development of the failing Gooch WMA. It was beyond my imagination that this group could be so inflexible in their thinking that they could not change their minds, especially since the fallacy of their stance had already been aired to its fullest extent by reputable natural resource professionals.

In retrospect, I see that ignorance could probably be breached but not hubris. We had thought that good, sensible thinking could change any receptive mind, but the malady continued.

A Significant Meeting with Myers, 1988

In fact, the matter should have been resolved two years before my meeting with Campbell. Director Myers wanted the problems formally addressed. Dan Sherry, the TWRA environmentalist, and Dan Eagar, an environmentalist from TDEC, spearheaded this effort. A committee that included me and more than a dozen other natural resources professionals was called together to answer questions about the compatibility, feasibility, and values of the Tigrett plan.[2] The task of the committee was to address three options: (1) implementing the full project (cleaning up swamped-out wetlands); (2) carrying out a light-touch version (cleaning out old meanders, improving hydrology without levees and ditches); or (3) undertaking no project at all (letting the floodplain recover on its own).[3] Myers included the stream-obstruction-removal guidelines that Chester McConnell and others had promoted as well as our river restoration concepts. A thoughtful man, Myers never lacked energy to do what he felt was right. But the issue had reached a point that all hope for recovering the rivers might be lost.

It was around this time—the spring of 1988—that I began to better appreciate what was probably one of the greatest struggles Myers had faced in his career. The clash between the litigants in the Akers lawsuit over the WTTP and our philosophy of management was continuing to heat up. The litigants' position that the mitigation land be accepted and developed—a stand that would revive the corps's channelization work—had become the most controversial subject in the natural resource and conservation communities. Myers remained largely silent—no doubt because the conflict weighed so heavily upon him and because he understood the uncertainty of the outcome only too well. Yet, he still seemed committed to our cause; otherwise he would not have set up so many of the critical meetings in which we were allowed to have our say. Not once had he instructed us to back off. My own take was that he wanted to get the mitigation land first—in the hope that somehow, in some way, a compromise could later be reached that would stop channelization and allow us to manage the mitigation land properly.

In our committee's meeting with Myers in May 1988, we hoped to get him off the fence. He wanted candor, and that's what he got. The question was this: should he accept the litigants' position—one that was supported by Ron Fox, his field director—or should he listen to his staff and representatives of other agencies? If he had hoped that the committee would find in favor of the WTTP, Akers, and Fox, he would be disappointed.

The meeting gave us a chance to address some previously neglected issues, such as the state of the fisheries in the area, something that had long concerned me. Wildlife management in the Obion–Forked Deer watershed had always emphasized waterfowl, particularly ducks, but surveys conducted several years before our meeting showed that, even with the degraded state of the river system, the demand for fishing had significantly exceeded that of waterfowl hunters. The TWRA had been criticized in the past for an allegedly "elitist" attitude that favored waterfowl projects over those for fishing. Thus, one of the committee's conclusions was that "the fisheries issue should be a major factor when considering our position on this project."[4]

But the primary debate at the meeting was whether to accept continuation of the WTTP and channelization, or whether to intervene and insist that the corps adopt a "light-touch" method that would remove blockages—sediments, logs, and other debris—from the river channels and thus allow them to reach stability on their own, something that might otherwise take a hundred years. Everyone on the committee—with the exception of Ron Fox—agreed that the "light-touch" method was best. It was not that Fox believed the light-touch methods were bad for the river but that insisting on this approach with the corps would muddle the process and foul the acquisition of the mitigation land. Fox's argument, as summarized in a report on the meeting that was submitted to Gary Myers, was that "most of the impacts [by the WTTP] have already been incurred and a mitigation package is assured." He contended that "tampering with the concept of abandonment of the project would jeopardize the only compensation we would get from a largely completed project."[5]

Also at issue was how the lands should be managed. Supporters of the Akers lawsuit advocated methods that were similar to what had been done at Gooch WMA—which, to those of us in the field, was a prime example of how waterfowl areas should *not* be developed. On this issue, too, Ron Fox believed that changing the plan from what had been recommended by Akers's supporters ran the risk of losing funding for the mitigation lands.

Those attending the meeting from our region recommended abandonment of the WTTP and keeping the thirteen thousand acres of mitigation land already acquired. As the report on the meeting pointed out,

> One of the major concerns expressed by almost everyone lies with the fact that the mitigation lands will not likely be developed to anything close to the degree called for in the mitigation plan. Even if developed [it] would not interact with the river as a natural wetland. . . . Other public interest values [other than ducks or fish] would be lost. . . . It is clear that as a stream recovers from channelization, the fisheries resources also recover. Undisturbed and given time, this recovery becomes complete. On the other hand, if the proposed project [WTTP] takes place, we must assume that the same channelization philosophy that has directed maintenance of the basin for the past 70 years still survives and will for the foreseeable future.[6]

"Tennessee," the report concluded, "would be better off with a reasonably mitigated stopped project than the present project fully mitigated."[7] This was clearly not likely to go over well with the proponents of the WTTP and the Akers lawsuit. The report was never made public, but I am sure that Director Myers struggled with its implications for many long months afterward.

The Role of Sportsmen in the Debate

Strangely, Akers's original lawsuit during the early 1970s lauded the restoration of the old river as the answer to channelization. Whether or not the plaintiffs understood anything about the need for the original river or positive hydrology is uncertain. Opinions, however, changed about the need for natural meanders once the court ruled in favor of the thirty-two thousand acres of mitigation land and the corps was allowed to continue digging the "Big Ditch." Now, to threaten the levees, ditches, and water-control structures intended in the mitigation plan was enough for the litigants and the Tennessee Conservation League to condemn us. Their position became a wall, heavily fortified. At this point, it appeared hopeless that recovery of the free-flowing river would ever happen. This feeling, however, had not reached local sportsmen in Dyer County.

The TCL, contrary to the popular belief, did not necessarily speak for the sportsmen of the state, especially the sportsmen of Dyer County, on these troublesome issues. "The state has a lot invested in this strip right through here [Tigrett WMA]," Ralph Lawson, a local lawyer and duck hunter, said in defense of the Tigrett plan. Video producer Doug Viar and I had just met Lawson on Beaver Road to interview him in 1991 in preparation for a video we were developing. Lawson was in a duck-hunting club smack in the middle of the Tigrett floodplain. "When you had a lot of green timber, you had a lot of food for ducks," he said. "Of course, if you can get the water off, you'll have a lot of other game, too. It will enhance the value of our property, if you can get this [river] system like the Hatchie."[8] Fire Chief Billy Taylor of Dyersburg still echoed this sentiment two years later. "I am here to see that the TWRA restores this wildlife area," he said at a 1993 meeting at the Dyer County Courthouse. "If it takes until doomsday, we will still be here." The sportsmen of Dyer County were not a shy lot.

Surprisingly, the sportsmen from outside of Dyer County had little to say about the Tigrett plan. This was probably because they had no means of gaining constant information about the issues; thus, they had trouble understanding them. Many on the staffs of TWRA and other agencies involved in the river restoration issues feared backlash from this silent majority. Campbell was right about one thing: there was a segment of duck hunters who traditionally hunted the old swamps and had built duck blinds here. For these hunters, the threat of draining their duck holes was enough to convince them to condemn river restoration, no matter what the benefits were or who else would be affected. It was an attitude too common in the country.

No one but a hunter can appreciate the attachment and sentiment one can feel about an old hunting ground. But not all hunters fit the same mold. There are distinct categories of duck hunter: the "dyed in the wool" hunter and the "if it flies, it dies" hunter. Those in the former category survive from season to season only to see another decoying flight of mallards. These hunters carry the novices to the duck holes to teach them duck-hunting skills, and vow—at least from an intellectual perspective, and often a heartfelt perspective—to uphold the law, the tradition, and the conservation of waterfowl. If they are in fact sincere, they are the "true" sportsmen, a valuable asset to the future of waterfowl and waterfowl hunting. One day they will be the major voice that helps to straighten out this debacle over which way is best: the anti-river, or native river; the counterfeit river, or the true and most beneficial river.

This leaves us with the "if it flies, it dies" duck hunters. These folks may wear caps bearing the logo "Waterfowl for Tomorrow," but for them it means, "Only if I am there to kill the last one." There is no hope for these hunters—we pray they are a minority—who would kill the last duck, the last sparrow, or the last redheaded woodpecker to gratify their quest for feathers on the ground. If they have their way, their children at an early age could well see the last duck. Such hunters will not support river restoration if their duck holes are threatened in the slightest. Like other exploiters of the river, they are the enemy at their own gate. Ultimately, if their attitude prevails, they will carry the greatest responsibility for eliminating the river's natural resources—the ducks, the hunting, and the native wetlands and all the benefits that go with them. I did not intentionally work for these hunters.

With confidence, however, I will boldly predict that the hunters and landowners with a sense of conservation ethics will ultimately have had enough, that they will be heard and prevail. To be sure, the "if it flies, it dies" crowd will still be around, but only to enjoy the outcome.

Dyer County Sportsmen and the Tigrett Restoration

Wilmer Elgin is a responsible Dyer County sportsman, who, at more than ninety years of age (as of this writing), spoke most effusively about the river and wildlife at one of our meetings. "What good is a squirrel without a tree," he pointed out. He talked about the hordes of ducks that once came to the pin oak flats at Tigrett, which, he said, had more ducks than the government refuges. "Oak trees must have been more than ninety feet tall," he continued, "hardly could reach my arms halfway around one. I could take my grandkids and walk right across the bottoms to the river and fish. We'd pick up a bucket of pecans and hickory nuts on the way back, and it attracted ducks. I hope I live to see the river restored."[9] Testimonies of this sort began to regenerate interest in Tigrett.

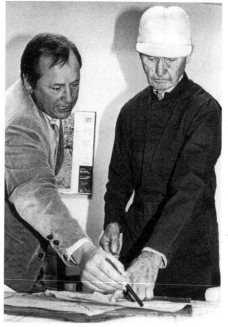

Wilmer Elgin (right), a longtime Dyer County sportsman, examines Tigrett restoration plans with the author. (Photo by Kathy Krone, courtesy of the *Dyersburg State Gazette*.)

Reviving the Dwindling Interest at Tigrett

Chief Billy Taylor waged a personal campaign against the TCL for its decision not to support the Tigrett project. Taylor's allies included Willie Reasons of the West Tennessee Wildlife Resource Association, the local sportsmen's club; Don Dills, the local state legislator; Joe Chriswell, who walked across the state for the conservation cause; Robert Harrell, a cartoonist; and Bob Dean, a local businessman and conservationist.

These men probably did more than anyone to gain support for both the restoration of Tigrett and the projects at Reelfoot Lake. Chief Taylor would not stop his campaign until he felt confident that Dyer County, the local politicians, and the wildlife commission were behind the Tigrett project. Unlike the Reelfoot community, Dyer County sportsmen had an organization. They also had grit and used it for the right reasons. They understood what had been lost at Reelfoot Lake, as well as on the Forked Deer River. They also knew the consequences of inaction. The people at Reelfoot were not short on grit either, nor were they slow-minded, but they generally did not organize in groups of more than three. This lack of organization hampered their will to do things they believed in. It cost them millions of dollars in lost funding opportunities for Reelfoot projects and the security of their hunting and fishing grounds. It also held back the management of the lake.

Since the Reelfoot Lake sportsmen failed to show up in support of the lake drawdown, Chief Taylor and his group filled the void. Because of their efforts, the TWRA Commission went on record in support of both the Tigrett and Reelfoot restoration projects. In general, the limited public opinion favored river restoration, as did some of the news coverage. Several papers—the *Dyersburg State Gazette,* the *Jackson Sun,* the *Memphis Commercial Appeal,* and the *Nashville Tennessean*—carried extensive articles about our efforts to restore Tigrett WMA and the rivers; all, one could conclude, broadly supported our efforts.

Nonetheless, the magnitude of what Tennesseans had lost by the destruction of the rivers could not be reached in these short articles. Thus, those not associated with the Dyer County supporters were rarely represented. At least we had the enthusiasm of the Dyer County sportsmen to encourage us.

The *Lost Rivers* Video

Even with the renewed interest, progress on the Tigrett restoration proposal seemed to take two steps backward for every step forward. It dragged on well into the 1990s. At one point, when it appeared that nothing more could be done, Director Myers stepped in. Knowing the dismal outlook, he called for a public relations strategy that included a video to reach the broader public. In February 1991, I drafted the first script for the video. Gary Cook, the new assistant regional manager (later the Region 1 manager), assisted with the edit. The draft script was finished by the end of the month. Myers authorized ten thousand dollars to develop a ten-minute video.

Dr. James Byford of the University of Tennessee at Martin met me at my house in Dyersburg soon after the script was prepared to discuss the river projects and to learn about the intended video. Jim encouraged me to continue with our commitment to the rivers in spite of the lagging interest from other agencies. He offered to assist in the development of the video before heading back to Martin. He must have started to work on the project on his way home. A day or two later, I received a song he had written: "A River That No One Loves." The song spelled out the dilemma of the rivers and the need to restore them. Later, we edited the song and changed the title to "Ghost River," after that segment of the Wolf River where Chester McConnell and I had our debate about the swamps.

The rivers gained an important ally in Jim Byford. His earlier background as a wildlife biology professor at UT–Knoxville had already brought him respect in the conservation community. His present position as Dean of Agriculture and Applied Sciences at UT–Martin drew equal respect from the farming community. He could not have come along at a better time. From that point on, Dr. Byford played a pivotal role in all the river management issues in West Tennessee. Soon we set to work on the video production and the song. Dr. Byford, accompanying himself on the guitar, provided the theme and background music for the production.

The video, called *Lost Rivers,* took on a life of its own. Doug Viar of D'Lomar Productions (based in Jackson, Tennessee) became the producer. Fortunately, Viar was a conservationist at heart, and through his skill and innovations, the video was stretched to twenty minutes instead of ten. The production contrasted the naturalness of the Hatchie River with the channeled Obion–Forked Deer rivers and thus clarified the techniques proposed for the restoration. Numerous conservationists and informed citizens provided on-camera interviews in support of the rivers. The first copy of *Lost Rivers* was produced in June 1992. More than four hundred copies were distributed to schools, the news media, individuals, and others.

The *Lost Rivers* video was a good stride in the right direction, but much more remains to be done to fully awaken the public. It will take a joint endeavor by all agencies— TWRA, TDEC, USFWS, and more than a dozen others—to generate a steady flow of information to the public about the necessity of restoring the rivers and their wetlands. This much-needed cooperative effort has yet to be realized.

Research and Technical Interest at Tigrett

In the meantime, though, the public relations push that began with the *Lost Rivers* video has been accompanied by an impressive array of supportive scientific research. George F. Smith's 1974 doctoral dissertation regarding the WTTP paved the way for a balanced view of channelization and the value of the Obion and Forked Deer rivers.[10] Smith described well the possible alternatives involved in weighing and assessing environmental quality. However, his was a very technical work that challenged the interpretive skills of even trained biologists, and the one important thing it did not include was a clear assessment of the generous benefits that come from native rivers, as opposed to the considerable liabilities caused by mismanagement. While Smith's work enlightened us, it left us with few options other than to continue evaluating environmental quality in terms of dollars, and dollar values alone will always fall short of measuring the true worth of natural resources to society.

Even so, dollars can still tell us quite a bit, as other researchers have shown. Sandra Postel and Brian Richter ably summarize such eye-opening research in their book *Rivers for Life: Managing Water for People and Nature*. Of one study, they write:.

> With the floodplains increasingly threatened by existing and proposed dams and irrigation schemes upstream, researchers Edward Barbier and Julian Thompson evaluated the economic benefits of direct uses of the floodplain—specifically for agriculture, fuelwood, and fishing—and compared these with the economic benefits of the irrigation projects. They found that the net economic benefits provided by use of the natural floodplains exceeded those of the irrigation project by more than sixty-fold (analyzed over time periods of thirty and fifty years). . . . Had Barbier and Thompson been able to estimate habitat supply, groundwater recharge, and other critical ecosystem benefits provided by the intact floodplain, the disparity in values would have been even greater.[11]

And of another study, Postel and Richter note:

> During the mid-nineties, University of Vermont researcher Robert Costanza and a team of ecologists and economists assessed the current economic value of seventeen ecosystem services for sixteen biomes. . . . [The roles of freshwater swamps and floodplains] in restoring and retaining water, mitigating floods, and breaking down pollutants emerged as particularly valuable. Rivers and lakes, which the research team assessed together, were valued at $8,500 per hectare per year, with the greatest value attributed to their roles in regulating hydrological cycle and providing water supplies.[12]

Scientists and engineers have no idea how to re-create many of these complex natural processes. While we live with substitutes, Postel and Richter point out, these are imperfect, and this gives us more reason to be extremely cautious when it comes to disrupting natural ecosystems like rivers and their wetlands.

In Tennessee, state universities were slow to follow up on the pioneering work by George Smith, but when the research did begin to emerge, we were ready to support it. No one had attempted before to restore tributaries of the Mississippi River. Until we learned about the Kissimmee River project in Florida, the only relevant research we knew about involved stream-obstruction removal methods, such as those used on the Wolf and Hatchie rivers and on the Obion and Forked Deer by the West Tennessee Basin Authority. None of this work, however, could match the magnitude of what we hoped the demonstration project at Tigrett would do for West Tennessee rivers.

Professor Dan Combs from Tennessee Tech reminded me that the Corps of Engineers had $22 million to undertake wetland research, and he was eager to start such projects. Unfortunately, that funding never materialized. Meanwhile, however, Professor David Bueler, a wetland specialist at UT–Knoxville, solicited an EPA grant to conduct research at Tigrett. Professors Reza Peseshki, Jack Grubaugh, and others from the University of Memphis successfully completed some of this work during the 1990s. Several of their graduate students also conducted wetland research at the WMA, among them James M.

"Mike" Oliver, who later became a wetland forester with Ducks Unlimited. Examining the soil hydrology at Tigrett, Oliver's 1995 study documented how important the interactions between native vegetation and hydric soil types are as they relate to creating healthy bottomland hardwood ecosystems.[13]

"People pay very little attention to the hydric condition of soils before using it for dwellings, agriculture, highways, or anything else," Oliver wrote. "Native vegetation, especially a tree, can be very particular about where it will survive and thrive. Thousands of dollars are often spent uselessly reforesting bottomland with trees that cannot grow on the sites they select."[14]

W. J. "Ben" Wolfe, and Timothy H. Diehl—hydrologists with the U.S. Geological Survey—conducted contract investigations on the Tigrett floodplain. Their objective was to determine sediment rates and surface-water flow paths in the area. Their final report was not completed until 1993. Many of their conclusions reinforced several points in our Tigrett plan. One was that the breaching of levees and spoil banks along the main ditch would increase the hydraulic continuity between the ditch and the floodplain. Also, they showed that, to concentrate flow in the flood plain, levees should be breached to convey water from the ditch to the floodplain at the upstream end of the floodplain. This would route water through flow paths in the floodplain and return the flow back to the ditch at the downstream end. To attain the maximum effectiveness, they pointed out, breaches should be located near existing flow paths. This was to say that we should follow the natural drains in the floodplain.[15] Diehl also speculated that at some sites along the main river channel, a sand horizon was as little as three feet below the bed of the channel. He expressed concern that a catastrophic collapse of the channel was possible in the vicinity of Tigrett if the channel were dug deep into the sand layer, or if head cutting penetrated this unstable layer of soil.[16]

This research in the early 1990s led TWRA to hope that more funding sources could be found and that the research could be continued. We conducted field trips throughout the summers of the early 1990s with university researchers and other experts in the wetland sciences.

As part of the research effort, a southern bottomland hardwood consortium was held at Dyersburg to discuss forested wetlands. Tigrett was the main target of discussion. Ken Arney, our chief of forestry, and I conducted a field trip for this group that included an important group of scientists. Arney also arranged for Earl Grissinger, Charlie Cooper, and Dan Marion from the U.S. Forest Service's hydrology laboratory at Oxford, Mississippi, to tour the project and make recommendations on proposals to reforest portions of Tigrett after restoration began. These men had spent their careers conducting research on rivers, and their opinions were highly respected. Like others, they were somewhat taken aback with our low-budget approach to restoring a river. Nonetheless, they encouraged us.

A more comprehensive wetland conference resulted from this interest in the West Tennessee tributary projects, probably with the encouragement of Jim Byford. Our work in the rivers was one of the featured topics. The conference, titled "Wetlands—Common Ground," was held on January 23–25, 1991, at the West Tennessee Experiment Station in Jackson, Tennessee. The gathering covered more than a dozen topics that encouraged a broad view of the values and management of wetlands. From all these efforts, we hoped a scientific consensus about river restoration would emerge.

Meetings with the public, various technical groups, and the Nashville staff continued on a regular basis from 1991 through 1992. We met locally with the farmers at Stokes Creek several times afterwards at Walton Guinn's cotton gin, the West Tennessee Wildlife Resources Association sportsmen's club in Dyersburg, and the Dyer and Crockett County commissions. Presentations were also given for the TWRA commission meetings, the Sierra Club in Memphis, the Dyer County Civitan and Rotary clubs, and numerous others.

Although the wildlife area managers and I had once given up on Tigrett, it was encouraging to see that a genuine interest in the fate of the rivers and their wetlands was now alive and well, at least among natural resource conservationists. The task ahead was to educate the broader public. We now believed that possibilities still existed.

Coping with the Tigrett Swamps

The period from the mid-1980s through the mid-1990s was not a dull period for any of the projects in the northwest territories. Even as our hopes for restoring the Tigrett floodplains went up and down, those concerns did not slow the demands at Reelfoot Lake, or the routine business of managing our other wildlife areas.

The Tennessee General Assembly created a new mechanism, known simply as the Wildlife Fund, to purchase wetlands during this time. Thanks to the energetic efforts of lobbyist and TWRA commissioner Tom Hensley, the enabling legislation was passed in 1986, and through a dedicated tax, the state was able to purchase nearly sixty thousand acres of additional wetlands for public use. These included nearly twenty thousand acres in small tracts within the Obion–Forked Deer watershed; thus, the amount of TWRA-managed land in the northwest territories nearly doubled. The new land was often purchased without our being able to evaluate it or produce management plans for it. Joe Hopper, the TWRA's acquisition biologist, had little time in which to act. Since the TWRA had been deprived of new lands for so many years, every acre seemed precious, and the main thing was to acquire the land quickly and then worry about it later.

These tracts, of course, were nearly always located within the river floodplains and had become available for purchase only because they were too wet to farm. To my regret, most had no buffer zones to make them complete: any acre that could be farmed was usually excluded from the sale. But the value of buffers was a moot point when sportsmen were desperate for places in which to hunt and fish. The new lands gave us hope that we had reached a point where some extraordinary progress could be made, something Myers had promised. These could be the years that the waterfowl managers in my district had waited to see. For the first time, the loss of wetlands in our area seemed to have bottomed out; perhaps now recovery could begin.

Our mission was so clear in our minds that we moved almost in unison, but the new responsibilities that came with additional land purchases began to take its toll on my staff.

Ralph Gray, who had earlier transferred as the manager of Reelfoot Lake, had the responsibilities for lands managed on both the Obion and the Forked Deer rivers. In June 1992 these responsibilities were split, and Alan Peterson took over lands in the Forked Deer and the lower Obion River. Six months later, Carl Wirwa replaced Peterson, who was transferred to the staff position of information and education coordinator. This period was a time of almost constant adaptation and transition.

Carl was an excellent choice. Formerly the assistant supervisor of Area 11 law enforcement, he was familiar with Tigrett and wasted no time in catching up to the fast pace Peterson had set for him. He understood the river crisis just as we did, and he eventually became widely known for his dedication to waterfowl and wetland management. Carl and I spent endless hours and many days organizing and trying to determine the minor hydraulic features in Tigrett after Ralph Gray and I completed the initial reconnaissance. Our plan for restoring Tigrett, now going into its eighth year, seemed no closer to implementation. Even with new prospects and support from the governor's office, we were still in wait-and-see mode.

What could be salvaged? We were not sure. Not much was left of our original plan, but the Dyer County sportsmen had not given up and neither would we. By 1992 Tigrett WMA encompassed nearly seven thousand acres, as some three thousand acres of new land

Carl Wirwa, TWRA wildlife manager in charge of Tigrett
and other projects along the Forked Deer and Lower Obion
rivers, played an important role in promoting the restoration
of the rivers. (Photo by Jim Johnson.)

This Tigrett WMA cypress swamp was experimentally drained, revealing odd adventitious roots extending from the swollen buttresses on the trees—a sign of stress from stagnant water. (Photo by Jim Johnson.)

had been purchased since the plan for the area was formatted a decade earlier. Boundary lines were missing throughout the south sections, and these required remarking and posting. Carl and his wildlife technician, Bobby Norris, had this responsibility. The swampy conditions sometimes made an accurate survey almost impossible. Traveling through a natural swamp is not as difficult as one might think, but artificial swamps can hardly be described in words. Ordinarily, natural swamps are dry during late summer and fall; thus their bottoms are firm and negotiable. A living canopy of trees, like cypress, also keeps the undergrowth shaded and subdued. Artificial swamps, on the other hand, can be miserable places with soft, muddy bottoms. Depending on the age of the swamp, fully grown trees usually stand dead or lie prostrate at the bottom. Dead woody vegetation, cat briars, elbow bushes, willows, water maple, grass, vines, southern smartweeds, and lily pads create an entangled wilderness. Add a one-to-four-foot layer of organic mud, stagnant water of depths from an inch to three feet, and a generous number of cottonmouth snakes, and the swamp can intimidate even the crustiest "River Rat."

The property lines on our tract maps ignored the inaccessible swamps for reasons we discovered just before Carl assumed his new position as manager. Max Billingsly, a retired military colonel with a surveying business in Jackson, was ultimately contracted to resurvey some of these lost boundaries. It was not a survey he would soon forget. With a state-of-the-art global positioning system, satellite imagery, and other tools of his trade, Colonel Billingsly took on the task, but in a short time he realized his mistake: he had underestimated his bid for the cost of the survey. To his good fortune, we understood his dilemma. Before long, he told us that he needed more financing or he would be working at a loss. We helped him justify the request, and he was able to continue. I doubt whether he made a penny of profit from the job.

The need to hire Billingsly had became clear during one of our own attempts to locate some of those unclear boundaries. The corps, which had originally acquired the land and turned it over to us as part of the mitigation agreement, was asked to assist in this effort, and in October 1992, it sent its land buyer, Max Crowe, and survey contractor inspector, George Royal, to the area.

The boundary lines were well marked on the Eaton bluffs above the swamp, but after that, the lines vanished. Walking even well-marked boundaries on the high ground could be gruesome—cat briars, thick underbrush, and a thick layer of leaves were nearly knee-deep under the scattered oaks along the marked portion of the line. Hardly had the four of us—Crowe, Royal, Alan Peterson, and me—started down the first line before we encountered two large cottonmouths coiled among the leaves. I moved them aside with a forked stick. After attempting a defiant strike or two, the snakes slithered toward the swamp.

Shrugging off this encounter with some uneasy jokes, we continued our task—but only for a couple of yards. Two more cottonmouths appeared in our path. Nervous from our approach, they quickly squirmed beneath the leaf litter and disappeared. It caused us to wonder about how many other deadly snakes we had stepped on or over. George did not hide his anxiety and sent us ahead. Soon enough it became obvious that the marked boundary line stopped abruptly at the edge of the swamp.

After considerable effort wading chest-deep in the swamps looking for more marked trees, we gave up. It became obvious that the corps's original surveyor had no stomach for such dank and forbidding places and had not mentioned the incomplete survey. The corps had then turned over the property to the state without noting this infraction of duty. Thus, with the job yet to be finished, the state contracted Colonel Billingsly. This doubled the government's bill.

Part 3

Facing the Uncertain Future

Chapter 16

The Burial of the Rivers

By early 1993 the Tigrett project was a struggle, and I had become exasperated. What good was it, I asked myself, if the project could not demonstrate a river restoration? The TWRA would be worse off than if nothing had been done. No one would have any confidence that we could ever restore the rivers. Critics of the original plan had modified it in such a way that, if implemented, the Tigrett WMA would be virtually disconnected from the normal flow of the river, and this was no way to ensure a self-sustaining floodplain. Engineering proposals added by Don Porter, a member of the Tigrett Technical Committee, would elevate costs to a level our budget could not support. Reforestation by conventional methods—clearing and planting trees—was added to the plan; this was a preferred approach, to be sure, but also a very expensive one. And there were other hindrances. We did not have the easements needed to restore parts of the old river meanders on adjacent private land. Nor could we reach agreement on the type of equipment to use for the reforestation work. Also, I and a few others felt that we lacked the support we should have had from TWRA's Nashville staff. Most of these problems would take years to resolve.

On Sunday, January 31, 1993, I decided that enough was enough. What I was about to do might be unfair to our supporters in Dyer County, whose encouragement had given us the strength to persist this far. Yet, I knew that if Tigrett failed to recover to their expectations, we could lose them too. After all, they were probably as weary as we were.

I pondered my decision for a few more days. On February 3, I called Harold Hurst and informed him that the time had come to retract the 404 permit application. I recommended that we notify the corps's regulatory office, recall the application, and get on with other business. At the very least, we might learn if there really was support to continue a campaign to save the rivers. Harold agreed. I then called Gary Myers and told him of our decision. He also agreed but wanted to retain plans to work on Stokes Creek, where the farmers had been desperate for flood relief for a long time. To my thinking, there was hardly any way to effect this drainage properly unless the floodplain also drained properly (that is, in accordance with nature's way), especially where Stokes Creek entered the Tigrett floodplain. Myers had promised the farmers he would try to solve their problem, but the "fix" in the modified plan now called for a straight ditch that cut laterally across all the natural drains in its path. This approach had been tried in various places before—

including Stokes Creek—and had failed. But, still, Myers had made his promise: a lateral ditch it would be if that was all he could manage.

I agreed—reluctantly. The farmers needed help and would likely consider temporary help better than nothing. However, I did mention to Myers that the work would have to repeated again and again. It would never be sustainable until the river through Tigrett was also restored. Perhaps Myers agreed, perhaps not. He did not say.

I added that it would be a good idea for someone from his staff to call the regulatory office in Memphis and inform them of our decision. I thought he said, "Yes, I'll do it." Harold Hurst, I noted, had planned to be at the next staff meeting in Nashville to discuss things further if necessary. After my conversation with Myers was done, I called Harold and gave him the details. Then, I went to bed early and slept for a long time, well into daylight the next morning.

Despite this decision, through the spring and most of the summer, we remained engaged in presentations of the Tigrett project in Dyer County. At one meeting in March, sixty people showed up. Support for the plan was rising, and somehow I did not have the heart to tell them of our decision. The Dyer County sportsmen and conservationists were encouraging more meetings on Tigrett and contacting the media. They wanted the Tigrett plan to be front-page news.

Meanwhile, I had heard nothing about the Tigrett 404 being withdrawn. In fact, I was surprised when, on July 1, Randy Clark and Tim Davis of the corps's Memphis regulatory office came to my home in Dyersburg to talk. They were excited about the Tigrett 404 application, saying they had received more support for it than any they had previously processed. This included one petition with sixty names, as well as many letters of support. Puzzled, I asked them if they had heard from our Nashville office about the decision to withdraw the application. They had not. A few days later, Clark returned to Dyersburg with Eva Long, an Atlanta-based field representative for the Environmental Protection Agency. Both still assumed the Tigrett project was alive and well. But it was not.

On August 15, 1993, I was at the University of Memphis attending graduation ceremonies for my youngest daughter, Kara, when I received a telephone message from our Region 1 office: "Meet Harold Hurst in Nashville tomorrow, 1 p.m." Of course, "Nashville" meant TWRA headquarters. Somehow I knew it had to do with Tigrett, and it could only mean more trouble. But what? I was soon to learn that Tony Campbell of the Tennessee Conservation League had called for yet another meeting on the Tigrett Floodplain Restoration Plan, and Director Myers had accommodated his request. The meeting would include the USFWS environmental office representatives from Cookeville, among others. To say I was bewildered is an understatement. What more could be done to trample on that poor river bottom?

Meanwhile, a new demonstration model for river restoration—the West Tennessee Rivers Mission Plan—had emerged, and although it did not include Tigrett WMA, it embodied nearly all of the same principles as the Tigrett plan. I will have much more to say about the Mission Plan in chapter 18, but suffice it to note here that it was one of the few bright spots during an otherwise difficult time.

A Meeting to Kill the Tigrett Plan

The 1 p.m. meeting on August 17 began on time, and the damnation of the Tigrett Flood-plain Restoration Project was the only item on the agenda. Myers was not present. Ron Fox, the TWRA field director, opened by saying that we were there to resolve the differences of opinions over the 404 permit application for Tigrett. No one seemed to know that Myers and I had already discussed withdrawing the application, and Myers, apparently, had intended for me to contact the corps about it. In any event, the tone of Fox's opening remarks made it clear that the intent of the meeting was to make sure that the plan and the permit died. But Harold Hurst and I thought that this was already the case.

The group Harold and I met with was small. It included Lee Barclay of the U.S. Fish and Wildlife Environmental Branch and his team, Bob Bay and Doug Winfred. Tony Campbell, TCL director, and his board director and civil engineer advisor, Don Porter (also a member of the Tigrett Technical Committee), were also present. David McKinney, chief of TWRA's environmental division, and his assistant, Dan Sherry, represented TWRA's Nashville staff.

Ostensibly, TCL's objection to the Tigrett restoration plan was its failure to detail the engineering needed to accomplish its goals. This objection drew on the opinions of Porter, who was a civil engineer with the Tennessee Valley Authority. We certainly agreed that engineering assistance would be appropriate once the principles of the plan were determined, but we believed that civil engineering that overlooked certain key principles—that all parts of the river should work together in harmony for the benefit of both the ecosystem and the citizens of the state—had contributed to many of the problems we were up against. From our perspective, the engineering profession had no business taking the lead in determining how the natural resources of Tigrett should be managed.

The TCL also complained that we had departed from the original plan by proposing the use of an amphibious trackhoe (a dredging machine mounted on pontoons) instead of "mules and handwork" to clean out the old river meanders. We had proposed changing the equipment because the amphibious trackhoe was faster and more economical to operate. Also, it had been demonstrated on Mayfield Creek in Kentucky to do a good job while having a low environmental impact. To us, it was an appropriate choice. But to our critics, advocating the use of this machine put us in the same category as "channelizers like Richard Swaim and the corps." (Ironically, the TCL had supported the WTTP, with its gigantic "dinosaur" dredging equipment, in the compromise over the mitigation lands.)

The TCL representatives also disliked the proposed route of Stokes Creek and the restoration of the natural river meanders, which called for breaching the levee along the excavated main ditch at the old meander outlets. And they certainly did not want to see the artificial swamps drained. In truth, they liked nothing about the plan and wanted it nullified. TCL director Campbell also reminded me that he wanted the Dyersburg sportsmen off his back. Porter, meanwhile, claimed that we had not been cooperative, that we had twice refused his offer to rewrite the Tigrett plan in an "acceptable manner." There seemed always to be a conflict between the principles of civil engineering and those of natural resource management. I wondered why the two could not be united to everyone's satisfaction, but that marriage never happened.

In short, the meeting did not go well.

One thing that nearly everyone in attendance agreed on was that a lateral ditch should be dug to appease the drainage concerns of the Stokes Creek farmers. On this issue, mine was the lone dissenting opinion. Earlier, as I mentioned, I had agreed with Director Myers that this might be the only sort of relief we could offer the farmers, given the dim prospects for the rest of the Tigrett plan. But now I took the opportunity to voice my strong reservations about the ditch. Its lateral orientation was directly opposed to the natural drainage of the floodplain, and in my view, approving it would give the poor farmers only a false sense of security. It did not represent a genuine attempt to resolve the problem. But the prevailing opinion seemed to be this: so long as the farmers' nagging but legitimate complaints were quieted, why should anyone else care?

What the Tigrett manager and I wanted for the Stokes Creek farmers was sustained relief, something the lateral ditch would not accomplish. Like the rest of our critics' recommendations, the ditch was a stopgap measure, and stopgaps have a way of becoming permanent while leaving the underlying problem unsolved. In 1973, for instance, Harvey Bray, then director of the TWRA, responded to political pressure and flooding complaints from local farmers whose property adjoined the Gooch WMA. Bray authorized what was supposed to be a temporary, one-season lateral ditch crossing Gooch Unit B. That ditch—a twenty-foot-deep chasm stretching into the foothills and onto private land—is still there and now appears as a permanent blue line on the topographic maps. This rerouting of Dillard Creek to the main channel of the Obion was supposed to resolve the farmers' concerns, but in time the drainage deteriorated and flooding increased, leaving the hydrology of the floodplain chaotic and unstable. Ruptured levees and sediment deposition pose a horrendous maintenance problem each year. I could see that these circumstances would be repeated at Stokes Creek. The decision to dig the lateral ditch at that location was simply a ploy to sweep the problem under the rug.

There was no mention at the meeting that a decision had already been made to withdraw the 404 environmental application. Even before I had left for Nashville, the project seemed dead, and the meeting was just the last coffin nail.

Chief Taylor's Response

Chief Billy Taylor was not so easily discouraged. In fact, his determination would continue for another three years or so.

Chief Taylor called me late in the evening, after I returned to Dyersburg, for a report on the Nashville meeting. He was not pleased with the outcome; he was not content to lose the skirmish he had started. I reminded him that TCL director Campbell was displeased with pressures he received from citizens like himself in support of the Tigrett project. The chief was furious: "If he wants relief, he needs to change his attitude about the project. He ain't heard nothing yet, friends."

Taylor and Joe Criswell, a local conservationist, kept up their campaign to restore Tigrett by contacting politicians and the news media. Taylor, in fact, must have called every television station and newspaper he could think of and told his story about the plight of Tigrett. The river, he argued, must be restored for the good of the sportsmen, the landowners—indeed, for everyone. He assured anyone who would listen that the TCL did not speak for the sportsmen in Dyer County. The TCL director softened his complaints and tried to reassure Taylor and his supporters. Campbell brought up the idea of cutting a temporary ditch from Stokes and Eliza creeks to appease the farmers. He said that he could live with a compromise, provided that the swamps were not drained.

But Taylor wanted results and had no use for appeasement. "I've seen these meetings before," he told me. "If they think they are going to 'hoo-doo' us in on some kind of whitewash—well, they have another thing coming." Director Myers responded by organizing a "restoration team" to reevaluate the Tigrett Plan.

The Nagging Swamps Issue

Meanwhile, the Sierra Club in Memphis had taken a position in support of the artificial swamps and badgered the public to oppose the Tigrett plan. We tried to explain that these swamps were not the wildlife oases they imagined. Nevertheless, they persisted, and the disagreement went on for two or three years. Sierra Club spokesman Larry Smith, who was considered the Memphis chapter's river expert, called the Tigrett project "the best propaganda plan I've heard of" to drain wetlands, telling the *Dyersburg State Gazette:* "I'm in a battle here. I'm taking people out to show them what they are losing."[1]

Smith's quest to save the artificial swamps reached amateur ornithological groups, regional sportsmen, and others through specialized news outlets. "The Tigrett plan should be scrapped," he wrote in an article for the Audubon Society newsletter. "TWRA should not use the nice sounding words of 'river restoration' as an excuse to drain valuable, productive wetlands."[2] Smith even coauthored an otherwise good report on stream channelization and swamp formation in West Tennessee. The argument was that large swamps similar to the artificial swamps at Tigrett had existed before European settlers came, and now we were trying to drain them.[3] The *Mid South Hunting & Fishing News* in 1993 ran a sizeable story about Tigrett, and it quoted Smith: "One group of hunters I talked with said they fish all the time and reported crappie fishing to be good." Greg Alford of Atoka, Tennessee, also a member of the Sierra Club, was quoted as saying, "The hunters [said to be more that two dozen parties] who came out of the swamp while I was there said there were a lot of ducks. This plan essentially calls for the destruction of the existing wetland habitat." And Smith added: "If the plan gets funding and is carried out, the same thing could be done throughout the state on other wetlands."[4] How right I hoped he was on that point.

If this was a contest, they were probably the victors. No doubt they had confused the hunting public, and they had probably convinced many people that river restoration was the lame idea of incompetent wetland managers. Their influence on the sportsmen was evident in another *Mid South Hunting & Fishing News* article. "Unfortunately," author

Peter Schutt wrote, "some managers are more intent on doing what's politically correct to keep their jobs, or [to] advance a hidden agenda, than they are on doing what's right for the environment."[5] If only he had known how many "political agendas" we had disturbed in our face-to-face confrontations, Schutt's opinion might have been different. It shows how well-meaning conservationists can ride to death their own "personal agendas" at the expense of much more important conservation agendas.

However, some of the locals who had hunted in the river bottoms most of their lives did not buy into these arguments. In a letter to the *State Gazette*, James A. Mallard expressed his views about the Sierra Club's position:

> I strongly disagree with the statement one of the canoeists [Smith] gave to a *State Gazette* reporter, saying the current effort to bring the Forked Deer River back to its natural channel "is the best propaganda plan I've heard of for draining wetlands." My wish is that these people could have walked with me in the beautiful hardwood timberland that once straddled along both sides of the North Fork of the Forked Deer River, all the way

The channel of the Forked Deer River through Tigrett WMA shortly after it was dug. Although the flood plain may appear to be a healthy forest, 90 percent of the trees have died. The arrows denote the natural path of the river. (Photo by Jim Johnson.)

The Burial of the Rivers

from Dyersburg to Tatumville and beyond. . . . In those days, the so-called "new channel" would overflow in the winter, pouring water 10 to 15 inches deep over the pin oak flats. At daybreak, thousands and thousands of ducks would pour into the area to feed on the acorns. When the water receded, it would drain back into the channel. There was an abundance of wildlife. Squirrels fed in the oaks, hackberry, and hickory trees. When fall came, we could hear from our home near the river the baying of the hounds ringing through the night as hunters followed through the timber on the trail of a raccoon. Rabbits were plentiful along the outer edges of the woods, and quail were in the nearby fields. But the area is now covered with water and button bush, and it can't even be traveled even with waders. Today, it is a smelly swamp; the once-beautiful timber is dead, and the wildlife is gone. Also gone are the old river beds or ox-bows that were loaded with fish.[6]

James A. Mallard said it about as well as it could be said. A lot of the Dyer County residents felt the same way he did, including Chief Taylor and his friends.

The Restoration Team Gathers

The "restoration team" that Gary Myers put together in response to the concerns about the fate of the Tigrett plan gathered on September 10, 1993. The schedule included an afternoon tour of the Stokes Creek area after an early morning orientation meeting. Following this trip, Taylor had scheduled a 7 p.m. public meeting at the Dyer County Court House. Another field trip was scheduled the following day.

The team was a familiar group. Myers, Ron Fox, Harold Hurst, Dan Sherry, Alan Peterson, and I represented the TWRA. Don Porter and Tony Campbell represented the Tennessee Conservation League, while Doug Winford represented the U.S. Fish and Wildlife Service. Billy Taylor was there for the Dyer County sportsmen. In addition, Nancy Strawn, the Dyer County Republican chairman, brought with her Steve Wilson, a candidate for the U.S. Senate in the Republican primaries. Nancy had pledged her support for our river restoration efforts and Wilson's, if he was elected.[7]

Fox, Campbell, and Porter carried most of the discussion during the morning meeting. The focus was on how to appease the farmers at Stokes Creek; it had nothing to do with solving the root problems or the management of Tigrett. The sportsmen's needs would be mentioned only at the local public meetings held later. After lunch we went to Stokes Creek. Several attendees opted out of the arduous trip. The temperature was more than 90 degrees, and as we discovered upon our arrival, the deer flies were ravenous. TWRA commissioner Curtis King from Humbolt had joined us for the field trip, but he, along with Tony Campbell, chose to stay with the vehicles. Chief Taylor, however, had no intention of missing any opportunity that might restore the rivers. Part of his mission was not to let a bunch of bureaucrats overwhelm him and leave without their producing a legitimate plan for restoring his desecrated hunting grounds.

The construction of the proposed lateral ditch at Stokes Creek provided plenty to talk about, although neither Doug Winfred, the USFWS environmentalist, nor I had much to say. However, I doubted whether Winfred supported the ditch construction. It took the better part of two hours for Don Porter, the civil engineer, to outline his plan. Nothing had changed.

After dinner, we attended Chief Taylor's 7 p.m. courthouse meeting. Some of the group on the field trip at Tigrett, including Ron Fox and Commissioner King, did not attend. County Executives Jim Jerman and Don Dills, Tony Campbell, several Stokes Creek farmers, WTWRA representatives Willie Reasons and Jim Keathley, and Kathy Krone, the reporter for the *State Gazette,* were all there. The gathering was not as large as Taylor had planned, even though the room was fairly crowded. Chief Taylor had wanted to have folks lined up at the door and standing in the streets.

Chairing the meeting, Taylor had two messages: he wanted to see action on the Tigrett project, and he wanted to assure everyone that the people in his area would not forget why they came. "We will not go away until the project is completed," he said. Jim Jerman, the Crockett County executive, and Don Dills, former Dyer County executive and state representative, among other leaders, stood up to support Taylor. Campbell and others attempted to appease Taylor, but the fire chief did not waiver from his straightforward message.

The following day at about 9 a.m., the same group went back to Tigrett to continue the field trip. Joe Criswell, who became known as the "Walking Conservationist," joined the group. Carl Wirwa had boats available to take the group up some of the old river meanders. But Ron Fox, the TWRA field director, showed no interest in the meanders and stayed on the riverbank, while Tony Campbell did accompany us on this day. Around 4 p.m. the group gathered at a shady place well into the swamp on Harrison Road to summarize and conclude the trip and, where possible, decide what the recommendations to make for the Tigrett plan. The shade and a chance to rest did nothing to hold down a occasional flare of tempers or to change old familiar biases. Carl and I could not discuss the subject of Stokes Creek without also mentioning the restoration of the floodplain.

In the end, the Tigrett plan lost three thousand feet of restored meanders east of Stokes Creek; it lost the route of the natural channel to connect Eliza and Stokes Creek; and it lost the restoration of the old river meander through the Brown Eddy. It also lost the breaches along the mainline levee on the north side. Not much was left of the Tigrett plan to argue about.

The meeting adjourned in a fairly congenial spirit. It was the last of these meetings to discuss the Tigrett issues. At least, our differences had been aired and, I hoped, would not be soon forgotten. Campbell smiled and likened to me a portion of a mule's anatomy. I smiled and referred to his ancestry. We shook hands and departed. All and all, most felt the trip was worthwhile.

Chapter 17

Black Swamp

While the struggle over Tigrett dragged on through the late 1980s and early 1990s, an even larger issue—precisely how thirty-two thousand acres of mitigation land resulting from the WTTP lawsuit would be managed—loomed like a dark cloud over the hopes for river restoration and green river valleys. The litigants knew what they wanted—all of the land and as soon as possible. They wanted it to be like Gooch WMA or a little better. If it were possible, they would turn all thirty-two thousand acres into flooded cornfields—that was the ultimate idea. What was going through their minds? Surely they did not think that the entire river floodplain could be managed for ducks like Clark Akers's cornfield. We had an excuse—ignorance—for the way we had developed places like Gooch WMA twenty years earlier. But we now realized that the state could no longer afford this high-risk, inappropriate business. Nor could we bear the thought that one day the public would demand an apology, since we now knew that levees and water control based on channelization were part of the chokehold that had nearly squeezed the life from the rivers. And yet the plan to develop the mitigation lands had not changed since the settlement of the Akers lawsuit during the late 1970s.

All the Wrong Ideas

Other events along the Obion and Forked Deer had been going on during the Tigrett WMA crisis. One was a meeting, held in July 1991 about five or six miles east of the town of Kenton, to discuss the development of Black Swamp, a portion of the mitigation lands that had been awarded to the state by that time. Harold Hurst and I were instructed to meet Director Myers and Clark Akers at Hop-In Refuge, just across the Sidonia Road from Black Swamp. The group was waiting at the gate when we arrived. With them were two TWRA commissioners, a former commissioner, and Tony Campbell of the TCL. This was not the first time this swamp had been discussed, but it was the first time we had a meeting with the supporters of the WTTP lawsuit—that is, the litigants, their supporters, and members of the Tennessee Wildlife Resources Commission.

Black Swamp was a magnificent eight-hundred-acre natural swamp with an abundance of bottomland forest; its characteristics were probably unique to the Obion–Forked Deer complex. Though disturbed by the impact of excessive sediments and channelization, it had managed to survive with good vigor. Huge cypress trees, many 3.5 to 7.0 feet in diameter at chest level, and water tupelo near a climax stage were at the center of the swamp. Large cypress stumps from a harvest that had occurred at least fifty years earlier attested to the swamp's historic productivity. Oaks, hickory, and other hardwoods were on the higher ground at the outer edge of the swamp. Spring seeps fed the swamp during the dry seasons, providing another interesting, if not rare, feature. The swamp was ideal for numerous birds, small animals, furbearers, and creatures endemic to a cypress-tupelo swamp ecosystem.

Those virtues notwithstanding, however, Black Swamp was not good habitat for waterfowl; indeed, it had never been a favorite hunting area among the locals. It did not produce the right kind of food, for one thing. Cypress and water tupelo mast are not highly nutritious for waterfowl, nor are they preferred by most. Although wood ducks often feed on the mast, the swamp was, at best, a resting and roosting site for ducks, not a feeding ground. It had few oaks and other sources of favorite mallard food, so use of Black Swamp by those birds was rare. Thus, to think it could be good for waterfowl hunting showed a lack of understanding about the birds' habits. Nevertheless, the question of developing Black Swamp for just that purpose had smoldered for a decade; now it was on fire.

Issues like the development of Black Swamp had been around since the first tracts of mitigation land were acquired. When the WTTP was halted in 1978, fourteen thousand acres of the mitigation land had been purchased and eventually turned over to the state. After the purchase, the land involved in Black Swamp had been surveyed and laid out by the agency surveyors for development—but that development never happened. Bitter that the WTTP had ceased without the complete acquisition and development of the mitigation land, J. Clark Akers now wanted to see some "dirt turned," to see at least part of his dream become a reality. Thus, he and his supporters wanted to transform Black Swamp, a still-vital natural wetland, into an artificial swamp full of ducks. The ducks were fine, but the proposed development methods and the potential for a devastating outcome were deeply troubling.

Akers could not be denied credit for stopping the original intent of the WTTP, but he had accepted the wrong advice for the development of the mitigation lands, the Black Swamp being the first of these. Akers speculated that thousands of ducks could be congregated and hunted here once levees and water-control features were used to regulate water levels. His vision was that Black Swamp could, for example, be flooded as a green tree reservoir, or GTR. Like many other hunters, Akers had seen ducks gravitate to flooded green timber along the Obion River in the 1950s. Green tree reservoirs were famous for duck hunting in Arkansas, Missouri, southern Illinois, and other places. Akers saw no reason why Tennessee should not create something similar in the Obion floodplains. What he did not know was that most of the green tree reservoirs were also prized tracts of timberland that suffered serious and often irreversible damage from flooding.

Although we managers had pointed out the fallacy of such proposals, neither Akers nor his supporters changed their minds. In fact, they dug in their heels. What time, experience, and copious amounts of damning physical evidence had done for us did nothing for them. Whatever the reasons for their relentless pursuit, we saw that it could very well end up destroying about half of the hardwoods around the swamp and stress or kill the cypress-tupelo stands. To my staff and me, developing the swamp along the lines Akers proposed was to ignore sound wildlife-management principles. It made no sense.

Contrary to good advice, TWRA's leadership was in favor of the project, and this position dulled the spirits and confidence of those of us in the field, who saw a disaster in the offing. In our view, the most damaging effect of this wrongheaded position was on TWRA's reputation among its sister agencies, such as TDEC and the USFWS: to this community of professionals, the TWRA director's office appeared hypocritical, considering that we had spent two decades promoting floodplain restoration. The full embrace at Black Swamp of what were probably the most environmentally unsound and damaging methods ever used to manage waterfowl (or any other wildlife, for that matter) did our agency great harm, in my view and that of my managers.

The Black Swamp situation had started out innocently enough. A few hunters with means and waterfowl hunting land of their own were simply trying to do good things for ducks and duck hunting. Eventually, however, it turned into something personal. Vendettas developed, and so did political manipulation and self-interest. Maybe it was just plain hardheadedness. Being an avid duck hunter did not make one a competent wildlife manager. And now Black Swamp had become a bio-political time bomb ready to explode. None of this should have happened.

All that was needed for the Black Swamp, Akers believed, was a dam that would flood the timber throughout the area. (This method had "worked" at Tigrett WMA—until all the trees died from the summer inundations.) Akers had used levees and pumps in his own cornfields a few miles downstream, and he had bagged ducks. Why would the plan, he reasoned, be any different for Black Swamp, and why could it not be duplicated throughout the rest of the mitigation land?

Some of Akers's closest advisors were professional wildlife administrators and should have known better. Thousands of acres of defunct wildlife land and nearly a thousand miles of ruined river in the Obion–Forked Deer complex should have been enough to convince them that the plans for Black Swamp were a mistake. Arkansas, Missouri, Illinois, and a host of other places had already begun to rethink the use of GTRs. From such examples, we felt that if the manager believed that he could not afford to lose the trees, he would do well not to develop GTRs, especially directly in the path of the river. For his part, Akers was matter of fact about his views. Perhaps he was unaware of the downside, or perhaps he did not care. I prefer to think that he had listened to the wrong advice; he may have simply rejected anything except what he wanted to hear. One thing he did have was enough clout to convince his allies.

At the upstream end of the cypress-tupelo swamp, four hundred yards across Beech Ridge, was an area that had been the original Black Swamp—at some point, the name was arbitrarily transferred to the current location. The original Black Swamp was a perfect example of how not to manage a wetland. A large levee some ten feet high had held water in the swamp since the late 1960s. The large oaks in this 270-acre area had died and fallen. Nearly all of the remaining trees—cypress and water tupelos—were also dead or dying. It was a familiar scene all along the Obion and Forked Deer rivers. The dead and dying timber in this swamp—once a magnificent tract of bottomland hardwoods—should have at least concerned them, but it did not. It seemed as though wetlands were expendable, especially if they conflicted in some way with their use as hunting grounds.

One TWRA commissioner said as much during the July 1991 meeting. "I am not worried about anyone but my own people [the duck hunters in his area]," he boldly announced. "The others can worry about themselves."

The opportunity to revisit the subject with Akers and his allies would not be possible for several more years. Director Myers and I had a conciliatory discussion about the first meeting in January 1992. We also discussed the state of affairs for TWRA waterfowl projects in the entire Obion–Forked Deer river system. He knew that our arguments were sound, but I believed that it was time to concede that our opponents' minds would not change, that we should move on. But Myers said, "It is not over. Time has a way of changing minds—and people."

Myers tried to find ways to compromise. At Tigrett, he suggested, levees might be built around and across the bends of the river to create and isolate a few of these swamps. That way the development would be like an island, in which the floodwater could flow around the development and the floodplain would be least affected. The proponents of artificial swamps might agree with this idea. We had essentially followed this strategy at Hop-in Refuge. Low-level terraces were built there, more or less on contours. These structures worked well with native food crops and demonstrated that high levees were not needed to manage waterfowl. The low terraces did not compromise the river, and our objectives were met without an exorbitant cost, without harming our neighbors, and without creating artificial swamps.

Given a choice, we would have preferred not to use terraces or any other development within the boundaries of the annual floodplain. However, in this case and nearly all others, we had no choice. Land in the path of the river was what we were stuck with. Given these circumstances, terrace development was a fair and acceptable compromise—except for hunters afflicted with the "hot crops or nothing" syndrome. Corn, they thought, was the only food that attracted ducks.

The issue of the Black Swamp put Myers between, on the one hand, a powerful interest group bent on developing the swamp and, on the other, our advice against it. "Clark [Akers] has done enough," Myers reasoned for self-comfort. "Why can't we give up a thousand acres and let him do whatever he wants [in Black Swamp]?" But if he thought this would end the problem, he would soon find that he was wrong.

The Nashville Meeting and the Black Swamp Plan

Myers had not forgotten his promise to revisit the Black Swamp. In 1996 he finally set up a meeting with J. Clark Akers Jr. and his supporters. Myers still believed that there was substantial good will among all the parties, that everyone was in fact seeking the best possible way to manage the river resources for the people of the state.

There was ample time to rethink the old ways that had failed in the four years that followed our meeting at Hop-In Refuge. The West Tennessee Tributaries Mission Plan was in the making (see chapter 18), and the proponents for the development of Black Swamp knew about it. Some, in fact, were a part of it. Concepts in the plan promised vast changes in the status quo for managing rivers. Erosion control and the restoration of the river meanders formed its backbone, and it was the most hopeful sign for the Obion–Forked Deer system in a hundred years.

Myers's meeting was scheduled for February 22, 1996, at TWRA's Nashville headquarters. He remained optimistic that relevant facts and common sense would change people's minds. The topic of discussion was the management of waterfowl in the Obion River, but the unmentioned purpose was to show that implementing Akers's proposal to develop Black Swamp would likely result in an area like Tigrett—a dismal failure. Besides destroying a rare and healthy cypress-tupelo forest while not producing the fine hunting that Akers so wanted, it stood to destroy one of the greatest initiatives of the century for managing West Tennessee rivers—what we now knew as the Mission Plan.

In attendance were Akers, TWRA commissioner Tommy Akin (who brought along a friend), and TCL director Tony Campbell. From Myers's staff were Ron Fox (the TWRA field director), John Gregory, and Ed Warr. My staff was also present: Paul Brown, Ralph Gray, and Carl Wirwa. All of us sat around a fairly large table in the extra conference room. I had arranged for a slide presentation.

The intended agenda was for me to give the opening presentation and background on the status of waterfowl and management plans along the Obion River; discussions would follow. Myers apparently had not mentioned the agenda to Akers, and he took charge of the meeting even before Myers finished his introduction. From Akers's point of view, the meeting was his. It was typical of Akers to walk into a meeting, make his points, and leave. It seemed that this one would be no different. If so, my presentation would end before it started, and I might have no other chance to present the material Myers had requested.

"I just wanted to tell you what I would like to see on this land we have been talking about. Now, what I have at my farm. . . ." Akers began. Myers did not stop him but looked at me as if to say, "Honestly, this was not what I intended."

"Mr. Akers," I interrupted. "If you don't mind, would you first indulge a thirty-minute slide presentation? What I have to say might influence your proposals."

"Sure, go ahead," he politely agreed.

The slide presentation left no doubt as to the messages intended. Who could argue with the practically useless floodplains and the miles of dysfunctional river? Without a mention of Black Swamp, I gave an overview of past management and the failures at

Gooch, Tigrett, and numerous private waterfowl hunting clubs. I detailed the effects of the levees and why we were against using them. I talked about the lost timber—how the trees had been drowned and rotted by the artificial swamps and erased even as marginal wildlife habitat. Even if restoration began tomorrow, I pointed out, it would take at least a hundred years to produce a mature forest. I described the destructiveness of flooding to our facilities and crops, the high cost of maintenance, and so on. I even emphasized our obligation not to harm the function of the river or that of neighboring land. Finally, I reminded them that in the end we had done a damn poor job of managing waterfowl but a fairly good job of fooling the hunters and ourselves.

I argued that we could avoid the ongoing problems—and achieve much greater success—if we moved our operations a few feet above the annual floodplain and restored the bottomland to native waterfowl habitat. It could not only meet the urgent need for upland habitat, but also produce a wider river corridor that would help preserve the integrity of the river and minimize landowner problems from flooding. Waterfowl hunters would find increasing hunting opportunities. Other sportsmen—deer hunters, turkey hunters, squirrel hunters, quail hunters—would also realize new benefits.

Nothing I said mattered; no minds were changed. I might as well have been talking to swamp-dead trees. Akers was courteous, if not patient, but I could not say the same for the remainder of his supporters. It bored them stiff. Finally, Akers excused himself to visit the restroom. I lost my audience but not its interest—I never had that in the first place.

Packets supporting the presentation were distributed, but Campbell waved his off: "I've seen it all before." I suppose he had, since he had been central to the development of the WTTP—as well as the Mission Plan just a few short years earlier. Only now did I understand that the conflict was not about the future of waterfowl or hunting. Truthfully, it was not about wildlife resources but was instead some sort of political chess game. Being open-minded or attempting to do the right thing had nothing to do with it. Had any of Akers's supporters understood the intent of Myers's agenda, I doubt they would have attended the meeting.

After my presentation, Clark Akers promptly started with his proposal. "I know little about your work on other places along the river," Akers said as he stood up. "I only know about the hunting on my place and in this part of the river. What I want to see . . ." The first thing he proposed was the development of a thousand-acre GTR for Black Swamp, with levees crossing the floodplain to retain water on the green trees. "I don't know anything," he continued, "about all that Tigertail land you manage [Tigertail was land fifty miles or so downstream near the Mississippi River and had very little to do with the subject]; that's your business. I am only interested in the land in my area." The methods he proposed were the only way he knew to manage ducks. He noted that not a single public waterfowl area had been developed as called for in the mitigation plan. The state had received the land as a result of his efforts, and he wanted it managed according to the original plan—his plan—embodied in the conclusions of the 1970 lawsuit. In fact, he declared, the ideal plan could be found on his farm only a few miles downstream; it incorporated levees, water-control structures, corn and water, and so on. Certainly, it was a successful operation for him and his supporters. There was not even a duck blind. He, Fox, Campbell, and Myers, I am told, had hunted the flooded cornfields nearly every season with nothing more than a few decoys and a chair placed among the corn stalks for a blind. Their hunts were, by any sportsman's estimation, a duck hunter's dream.

Not one of our recommendations was mentioned.

The discussion also strangely avoided any mention of the new river restoration initiative from Governor Ned McWherter's office—the Mission Plan. I wondered where they had been for the last three years. Surely they could not help but consider the principles behind the plan in progress and their role in the governor's reformulation of the WTTP. Akers and Campbell had been part of nearly every committee involving the WTTP. Ignoring our advice was one thing, but to me, it seemed rather cavalier to ignore a growing public policy consensus coming from the governor's office. All of it would come into sharp focus during the remainder of 1996.

The effect of the meeting was chilling. From that day forward, we knew that Akers and his supporters could not be reached. Attempts to change their minds ceased; there was no mention of the subject between us again.

Events since 1996

The final chapters on Black Swamp cannot be concluded at this writing, for the debates still continue. But as this book neared its final editing, conciliatory news reached my desk.

In April 1997 TWRA had applied for Section 401 and 404 permits to construct an 860-foot-long levee across the Black Swamp drainage, in direct contradiction to the Mission Plan. TDEC denied the Section 401 permit, and then the political storm began.

At the insistence of Akers and TWRA assistant director Ron Fox, the agency requested new permits. Director Myers applied for the two new permits in February 2002. Politicians were contacted at numerous levels, including the assistant secretary of the Army for Civil Works (head of the Corps of Engineers) and Tennessee's governor at the time, Don Sundquist. TDEC technicians, resisting until the end, finally issued the required Section 401 certification permit. Following suit, the Memphis District Corps of Engineers issued a Section 404 dredge-and-fill permit. It appeared then that the Black Swamp project would now be constructed, despite objections from the WTT Steering Committee supporting the Mission Plan, conservationists, the Farm Bureau, and others. It had the potential to negate every sound principle known to good wetland management.

Chester McConnell of the Wildlife Management Institute, who was also a member of the WTT Steering Committee, challenged the issuance of the new permits. The Tennessee Environmental Council and the Public Employees for Environmental Responsibility joined McConnell and filed an appeal with the Tennessee Water Quality Board in August 2003. Phil Bredesen was now the governor, and when the Black Swamp controversy reached his office, he called for an independent review of the situation. Two highly respected wetland ecologists, Leigh Fredrickson of the University of Missouri and Sammy King of Louisiana

State University, accepted the job. Their report, which included a jointly written section as well as individual evaluations by each scientist, concluded that "implementation of the Black Swamp plan would best serve neither waterfowl hunting nor the wetland."[1] The long-debated question of whether the proposed development of this wetland was worthy or not had received an emphatic and authoritative answer: "No."

"My judgment," Fredrickson said in his portion of the report, "is that it is not prudent to develop Black Swamp into a green tree reservoir."[2] The joint report said, "The combination of flood frequency and plant community would not be well served by the development into a GTR. Enhancements for a more 'natural' hydroperiod might be implemented but a levee and water-control structures are not the answer."[3] Fredrickson clearly stated that the design would compromise rather than emulate a natural hydrologic condition in a river floodplain in west Tennessee."[4] He also cautioned that the overly political nature of the pressure being brought to bear on wetlands managers was likely to degrade Black Swamp: "By reading between the lines in the information provided it seems clear that such pressure is part of the issue with Black Swamp and should be of concern if there is a commitment to develop a habitat-based conservation strategy for Black Swamp."[5]

If the truth of the matter made any difference, Fredrickson and King's authoritative report should have given the doubters now all they needed to make the right decisions.

Since then, a permitting process that began with a meeting in 1996 appeared to end on December 22, 2004. TDEC attorney Devin Wells, in a letter to McConnell and the other appellants, advised "that TDEC requested, and TWRA agreed to relinquish the [Section 401] permit" required for developing Black Swamp. At the conclusion of the letter, Wells added, "We trust that fair, competent leadership will evolve in TDEC and TWRA to prevent future unnecessary, costly controversies such as occurred with the Black Swamp project."

Under the weight of the Fredrickson-King report, along with appeals from respected conservationists, there was now hope for common-sense management of river wetlands. Too many mistakes have been based on ignorance and politics, and the case of Black Swamp should be a lesson to all that we can ill afford to waste money, time, and valuable wetlands on frivolous decision-making. Had something like the Fredrickson-King report happened earlier, thousands of acres of wetlands in desperate need of proper management could have been thriving and benefiting the citizens of Tennessee.[6]

But if changing something as entrenched as tradition and as complex as a manufactured river is slow and grinding, changing minds is even more so. The sensible thinking embodied in the Fredrickson-King report has been ignored by the opposition, who have become more determined than ever to have things their way. If they do succeed, the development of Black Swamp—not to mention countless other detrimental projects involving the mitigation lands—could yet become a reality with disastrous results.

Chapter 18

Ben Smith's West Tennessee

Tributaries Mission Plan

From the 1970s to the early 1990s, the river projects and those at Reelfoot Lake—all of which happened concurrently—were so intense and so fast-paced that it was often difficult to distinguish one from the other. Together, they should have given us all the lessons we needed for the successful management of West Tennessee rivers and their wetlands. But at the time, it was hard to sort out all the issues—to reach around them, so to speak—and it was especially hard to communicate what was needed to the public at large. It was also difficult for state politicians and administrators to see the bigger picture because of the different political and geographic boundaries involved. We in the TWRA seemed to have no pause, no breathing room, no time to deal with both the practical concerns of wildlife management and the larger wetland issues. We yearned to observe, to question, and to discover the lands the state owned so that we could meet the challenges that arose as growing numbers of people sought to use such severely limited acreage. But time for such reflection and analysis was rare.

By the mid-1990s, more than a decade after Frank Zerfoss and I had addressed the Nashville TWRA staff, hope clashed constantly with apathy, and very little seemed to go our way. We still had no new policy for managing waterfowl or the rivers, nor could we control the process. Not much could be pointed to as real and tangible progress.

It was in 1992, at an interagency meeting in Nashville on the troubled Tigrett project, that I got my first inkling of what would become the West Tennessee Tributaries Mission Plan—a model for river restoration that I have alluded to at several earlier points in this story. Conspicuously quiet at that meeting was a stranger, Ben Smith. He was from the governor's office, I was told, and my first thought was, "Another public relations man." In my view, we did not need more peacekeepers; we needed a common-sense natural resource plan supported by good engineering, as well as a course of action. Smith, when he did speak, outlined some possibilities for organizing a group to address river restoration projects. More committees? I had seen many of these already and was openly skeptical.

Maybe he could get a consensus from this group, Smith said. His brief presentation was tolerated about as well as spilled coffee on the paperwork strewn across the conference table would have been. The attendees just wanted to get past it and proceed with the meeting.

Ben Smith, representing Governor Ned Ray McWherter's office, produced the popular "Mission Plan" for river restoration, only to see it fall prey to politics and ongoing litigation. (Photo courtesy of Ben Smith.)

At first glance, it might have seemed a little surprising that Governor Ned Ray McWherter had enough interest in the West Tennessee river projects to send his own representative to the meeting. But McWherter had every reason to be familiar with the issues: his farm and home place rested in the hills of Weakley County, well within the watershed of the Obion River. Furthermore, such a ruckus had been raised over the WTTP that he could not ignore it even if he had wanted to.

Ben Smith had firmly established his reputation a few years before by heading Governor Lamar Alexander's Safe Growth Team. A Democrat, McWherter had no problem pilfering talent from the Republicans. When McWherter commissioned Smith to resurrect and "resolve conflicts over the future direction of the WTTP," he took a step in the right direction. Smith would not disappoint him.

Smith made a good start by apprising the governor of the relative success of the Mayfield Creek restoration work in Kentucky. This channel-improvement project used an amphibious backhoe machine to restore the meanders of the creek. It floated in the middle of the stream. Unlike the conventional excavation machines, the amphibious backhoe disturbed the surrounding environment very little while improving the function of the creek. This was favorable to the environmental regulators who often saw this type of work as controversial. McWherter recognized the potential for the machine's use on the Obion and Forked Deer.

Eventually, Smith showed up in West Tennessee. He asked for a meeting at our office in the Jackson State Office Building for 8 a.m. on October 25, 1992. Harold Hurst, Dan Sherry, and I were asked to be present. Field director Ron Fox had been instrumental in setting up the meeting with Smith and insisted, however, that the meeting be held in Nashville, not Jackson. Ben agreed to a Nashville meeting but kept the meeting he had scheduled with us at the TWRA office in Jackson. One thing was certain from that gathering in Jackson: Ben had a clear mission, and the points he made were well received.[1]

We met for more than two hours, discussing ideas and strategies for the new plan. From the beginning, it seemed familiar to us and for good reason: it was close to our own ideas. As he outlined the concepts in his plan, Smith revealed that he knew every detail of the WTTP, the Tigrett plan, and the *Lost Rivers* video. Having done his homework, he was well informed about the effects of levees in the floodplains. Hurst and I were impressed and elated. Finally, someone from the state's executive branch was on the right track. At the end of the meeting, I gave Smith a copy of the Kissimmee River EIS, which detailed the gigantic river restoration project in Florida, explaining to him that the principles in the Kissimmee plan differed from the Tigrett floodplain restoration plan only in magnitude. He probably already knew this. In turn, he produced a document on the impact of levees that he had commissioned from researchers in the Civil Engineering Department at the University of Tennessee, Knoxville, in 1987.[2]

I was surprised to learn that such work had been done. It was rare to find studies that directly addressed the need to improve the native floodplain instead of creating artificial or futuristic models. The well-done study went to the core of the problem—uncontrolled levees. It noted that the state had no laws to control levees and that these structures, which were supposed to prevent flooding, actually increased it. The researchers' conclusions surprised very few of us, but among politicians and state agencies, discussing such things was taboo. No one wanted the avalanche of criticism from the agricultural industry that was bound to follow. It was because of this ostrich-like habit that nothing had been done to solve the problem. We could have used the findings of this study many times earlier had it been available to us. But Smith advised me that the copy of the report he had given me was for my private consumption. I asked why he wanted it kept quiet. "The timing has not been right," he said. If not now, I wondered, when would it be right?

As it happened, around this same time, Governor McWherter asked the Corps of Engineers to reactivate—and rethink—the WTTP. The goal was to "find an environmentally sensitive design which would reduce erosion, restore floodplain integrity, and improve water quality." Subsequently, the governor appointed the West Tennessee Tributaries Steering Committee "to assure that the many public and private interests potentially impacted by the WTTP were consulted in project reformulation" and "to develop a project reformulation concept responsive to today's conditions, to new opportunities, and to the desires of local landowners."[3] Ben Smith now had his committee and was placed in charge of it. Most of its twenty members had already been deeply involved in the river issues and represented almost every possible interest: the West Tennessee Tributaries Association, the Obion River Conservation League, the WTTP litigation plaintiffs, the Tennessee Association of Conservation Districts, the Tennessee Farm

Bureau, the Tennessee Forestry Association, the Tennessee Conservation League, and the Tennessee Environmental Council. It also included representatives of state agencies: the TWRA, the State Planning Office, the Obion–Forked Deer River Basin Agency, and the Tennessee Departments of Agriculture, Environment and Conservation, and Transportation. From the federal government were representatives of the Fish and Wildlife Service, Geological Survey, Soil Conservation Service, Corps of Engineers, and Environmental Protection Agency. All city and county governments within the river basins were represented as well.

Smith called the members of the steering committee together on May 19, 1993, at the Tennessee State Capitol in Nashville. He told them that their job was to develop a concept for a river restoration plan, one that would address problems and improvements for natural resource management the landowners faced in the eight counties affected. From the outset, the plan explicitly sought to benefit not only farmers and duck hunters but "all Tennesseans."[4] It was the first time a river-management plan of this magnitude included benefits to the general public among its goals.[5]

The committee, whose members rarely agreed on anything (several had served jointly as board members of the Obion River Basin Authority and on WTTP committees), clicked off the criteria for the Mission Plan. During the summer and fall of 1993, there were three more meetings to work on the plan. At the next meeting in Jackson that December, the committee firmed up the goals and objectives of the Mission Plan, seeking total commitment from the players involved. The plan focused on the North Fork and the Middle Fork, Forked Deer River, and encompassed the upper limits of the corps-authorized project, some thirty miles up the Middle Fork to Highway 54 near Alamo.

What emerged from this process was a conceptual plan that was built upon a "tributary building block" idea. The aim, as Smith described it in a 1994 presentation to the state's legislative study committee on West Tennessee natural resources, was to restore the river channel's meandering course "where this was feasible, along with a comprehensive watershed program to reduce the amounts of silting delivered to the river channel by erosion of farm fields, road banks and ditches, and some critical gullied areas." The plan was unambiguous in identifying what needed to be done. It sought, Smith said, "to divert flows from the dug canals into the restored meandering river sections" and to breach, remove, and modify parallel levees so as to reduce dependency upon them.[6]

Most of the tributaries had some or all of the problems Smith described. Prioritizing these streams was central to solving the restoration puzzle. Initially, thirteen smaller watersheds within the vast WTTP area were identified as those contributing the most to instability in the river system. From these thirteen sub-watersheds, two were selected as early action demonstration areas.

According to the plan's recommendations, meandering river channels were to be restored. In places where it had to be incorporated into the restored channel alignment, the straight-line, artificial main channel—the anti-river—would be "tamed" by using in-stream grade-control structures to reduce head-cutting. Weirs, or small dams, would be used to divert low-flows into the restored meandering channels. Bridge modifications and "rip-rap" stone (large chunks of limestone used for erosion control and stream-bank stabilization) would be employed to reduce channel scouring. Impoundments in the upper watersheds of the tributaries would be constructed to reduce sediment loads into the river, and frequently flooded fields would be reforested.

This was heartening news. Now the new conceptual plan would seek to restore "positive floodplain drainage." Finally, some of the artificial swamps would actually be drained. Native bottomland hardwoods could be restored; natural wetlands would return; and wildlife diversity and outdoor recreation opportunities would soon return. The plan was truly multifaceted, and I could see that it incorporated all the important features we had longed to see in the Tigrett restoration plan. The steering committee adopted all portions of the Mission Plan by consensus. The committee's procedures required votes on pertinent issues; either the agreements were unanimous, or the next topic on the agenda was not addressed. The committee members seemed to recognize that Governor McWherter had provided a unique opportunity to set things right.

Smith's plan was finished in April 1994. Entitled "A Mission Plan for Reformulation of the West Tennessee Tributaries Project," it subsequently became known simply as the Mission Plan. When Smith and other state agency leaders presented it to the corps's Memphis district engineer on April 7, they received a very positive response. News media coverage of the Mission Plan was also very favorable, including editorial endorsements from the *Nashville Tennessean* and the *Memphis Commercial Appeal*. Governor McWherter and Ben Smith were lauded for forging a compromise that had been elusive for decades.

Other endorsements for the corps to change directions and support the Mission Plan, even from politicians, soon followed. Senator Jim Sasser, Senator Harlan Matthews, and Congressman John Tanner actively supported the new initiative, but Tanner still seemed to be the only West Tennessee politician willing to speak loudly to the public. He had shown this resilience when he joined Milton Hamilton, a state senator at the time, in support of the Reelfoot Lake projects, and these were certainly risky and proof that he would take chances for the things he believed in despite the risk to localized political popularity.

"In the early 1980s, we had much less to work with," Ben Smith told me. But, he added, "things had improved in the last ten years."[7] One change, according to an April 15, 1994, editorial in the *Memphis Commercial Appeal,* was the governor's stance on the WTTP channelization work: "Once a strong supporter of the WTTP, McWherter became convinced that its 'channelization' techniques of widening and straightening streambeds were unnecessarily harmful, as charged by opponents." Another editorial, appearing a few days earlier in the *Nashville Tennessean,* also reflected a turn in public opinion toward the restoration of the rivers; it observed that a channelization project that had cost $42 million, yet was only 40 percent complete, "never succeeded in easing the flooding it promised." Referring to committee's unanimous decision on the plan, the *Tennessean* editorial also noted, "Richard Swaim, the director of the long-embattled Obion Forked–Deer Basin Authority, said it best: 'We arrived at a consensus and nobody killed anybody.'"[8]

All of this was good news for those who had fought diligently to turn around the course previously taken. Smith's plan had revealed another way of managing rivers, and it had begun to gel in the minds of the public.

Once the Mission Plan was endorsed and advanced by the state, it was time for the Corps of Engineers to begin its work. First, the corps had to evaluate the project demonstration sites to determine the plan's feasibility, to tally the public benefits, and to determine whether it had a sufficient "federal interest."

The controversial Stokes Creek project at Tigrett WMA was included as a demonstration project site, although the larger Tigrett plan was excluded. The corps wanted more conceptual design details on how the restoration of Stokes Creek and the Middle Fork project (another demonstration site included in the Mission Plan) would be implemented. Since such technical work was too tedious for the large steering committee, a technical committee, which Smith also chaired, was formed to advise the steering committee. Several meetings, including many field reviews of the project area, kept the group busy through the remainder of 1994 and on into 1995, and I was fortunate to guide it during this period. On these trips, I gained support from the local farmers and the Dyer County sportsmen for river restoration. Finally, the technical questions were answered, and the months of waiting were paying off. In April 1995 the Corps of Engineers announced that both demonstration areas—the Middle Fork project site and the Stokes Creek restoration—were feasible and eligible for corps funding.

When it was finished, Ben Smith's Mission Plan had accomplished a seemingly impossible task. As far as I was concerned, Smith had exceeded even his own expectations. We had reason to celebrate; nothing before had offered the potential to resolve the conflicts between diverging interests over the rivers. The Mission Plan was long overdue. To Ben Smith, his committee, and the supporting political entities, we were deeply grateful and indebted.

Chapter 19

The Rivers Held Hostage

In my most optimistic moments, I believed that what we had tried to do to improve natural resources and the rivers would eventually become contagious, that state agencies and the public would support the Mission Plan, and that it would spread to all of West Tennessee. Local farmers could see a half-century of frustration, flooding, and creeping, unabated encroachment on their family farms relieved from the bondage of channelization and the misuse of the rivers. Landowners who gained little or no profits from swamped-out bottomland would see their forests regenerated and their investments reinvigorated. Hunters and fishermen would almost immediately begin to see wildlife and fish resources recover and their recreational opportunities increase. An enormous burden—in the form of flood damage to highways, buildings, and other infrastructure—would be lifted from the state. The cost of flood insurance would be greatly reduced, and emergency management would decrease dramatically. The devastated countryside—long abused and made ugly—could begin to heal. Everyone would find solace in knowing that the state had not shirked its responsibilities but had prepared for generations yet unborn. The general public might not realize what they had gained for a few years, but inevitably they would. The plan would fail only if its principles were overly compromised.

I could even envision the Mission Plan becoming a model for stream restoration in states all along the Mississippi River, from Minnesota and North Dakota to Louisiana and Mississippi. In turn, this interstate alliance could become a powerful voice on behalf of river issues for the country at large.

We waited, but nothing happened. The push for reformulation of the WTTP weakened, and finally it began to collapse. By 1995 Governor Ned Ray McWherter had reached his term limits and was succeeded by Don Sundquist, a former West Tennessee congressman. Attitudes took a turn for the worse, and despite some initially promising signs, little encouragement would ultimately come from the governor's office. Nor would there be support in Congress.

Early Optimism

Under Governor Sundquist, Dodd Galbreath, the policy coordinator for the Tennessee Environmental Policy Office, replaced Ben Smith as the coordinator of the Mission Plan and the ailing WTTP Steering Committee. He also coordinated the Interagency Wetlands Committee and its technical working group, whose purposes were to advise the governor on the status of the state's wetlands and to develop goals and objectives for those troubled areas. Governor Sundquist stated several times that he was strongly behind the Mission Plan and the WTTP reformulation, and it was heartening news when he endorsed a 1996 report by the interagency committee. That report said in part: "In previous years, over 59 percent of Tennessee's original wetlands have been converted or substantially degraded. The focus of any effort to restore, enhance or create wetland is to first establish natural hydrology from which all other attributes in a wetland will arise."[1]

It was a statement that no Tennessean with the slightest interest in the future of river wetlands should forget. But it was also couched in a language that the people of the state probably would not understand. Nor were they likely to see it since it was not widely distributed or publicized. All things considered, some of us doubted the new governor's sincerity. We worried that powerful political forces, unaware or uncaring about the state's wetland interests, were about to entangle the Mission Plan in compromises that would dilute it beyond recognition. The governor's office needed to recognize the gravity of the situation, for sooner or later the state would be held accountable for the outcome, and whatever that turned out to be would be for the governor, whether Sundquist or his successors, to address.

The public also had to come to terms with this conflict, or be doomed to suffer more tax burdens and lost benefits. Sooner or later, they would either appreciate the work accomplished by Smith and Galbreath or miss the historic opportunity they had provided. There was also good reason for the Corps of Engineers to be involved in the Mission Plan; it had the financial means and expertise the state did not have. The corps needed a partnership with local residents to redeem its past practices, and this project was right for both, not to mention the state as a whole and, indeed, the nation. If the corps could refrain from its habit of repeating work that had already been done (as in, for example, the redundant studies of Reelfoot Lake), a real partnership and real progress could result.

Even when a White House proposal to limit corps projects to those that helped more than one state looked like it might threaten the corps's participation in the WTTP reformulation, Mission Plan supporters remained optimistic. As the *Dyersburg State Gazette* reported in the spring of 1995: "Dodd Galbreath . . . said he hopes the corps will be allowed to continue working on the WTTP even if the White House proposal is approved. The WTTP could be grandfathered because it is so old and because it will provide a model for successfully bringing a variety of interests together to create an environmentally acceptable project."[2]

We also encountered hopeful signs from various sectors of the public. The Farm Bureau, an insurance company with a large membership of farmers and landowners throughout the affected area, lent its support to the Mission Plan—no small thing, since this was an inherently cautious group. In a 1994 letter to Ben Smith, the company's director of public affairs, Julius Johnson, commented favorably on an early draft of the plan, saying that the bureau leadership was "very hopeful and excited about the possibilities of moving forward and protecting Tennessee's resources in an agreed upon process." Such support echoed what we had already heard from sportsmen and others in Dyer County. I recall that at one of our last meetings in Dyersburg, held on March 19, 1993, those in attendance were not just duck hunters; business people and community leaders were there as well. And what Carl Wirwa and I heard after our presentation was encouraging indeed. These Dyer County residents were well informed about the issues and they supported our proposals. They knew that the Forked Deer was not a river to be proud of, and in fact, they wondered why we had not already started the restoration work.

Encounters like these, small though they were, left me convinced that when the public knew what was at stake and what we hoped to accomplish, they would get behind us. Indeed, this feeling seemed to be borne out in 1997 when a survey conducted by Dodd Galbreath showed that 66 percent of the upstream farmers—those beyond Dyer County—were in favor of stream restoration.

Disappointment

But to the dismay of those who supported it, the momentum for the Mission Plan would finally grind to a halt.

As we discovered, congressional funding and support were not forthcoming, even though the corps still had some $60–70 million left in its original WTTP account. That money would go a long way toward river restoration—if it could be used. But those funds were in limbo. Because the WTTP had been stopped by the state's refusal to issue the required water-control permit, the corps could not use its leftover funds for a reformulated project without the approval of Congress.

Here is where Governor Sundquist and Congressman John Tanner might have stepped in and used their influence to get the funds authorized. Tanner, however, claimed that the differences among the various parties over conflicts such as the Stokes Creek issue, the Black Swamp development, and the disposition of the mitigation lands first had to be resolved. In a 1998 letter to TDEC Commissioner Milton H. Hamilton, Tanner wrote, "I do not think it will be appropriate or productive to approach the Appropriations Committee about funding for the West Tennessee Tributaries Project until the State, Mr. Akers, the Corps, and other parties to the Consent Decree, have made the necessary changes to the Consent Decree. When agreement is reached, I will work with the Appropriations Committee and with U.S. Sens. Bill Frist and Fred Thompson to get funds appropriated for the project."[3] Of course, Akers and his supporters refused to budge unless Congress showed "good faith" and authorized the funds necessary to purchase the coveted mitigation land. It was a standoff that ensured that nothing would be done until one side "blinked." Leadership from the governor's office was not forthcoming either.

As Ben Smith, recalling those events of the 1990s, told me, "Sundquist . . . missed an opportunity to free the Mission Plan from the outdated constraints in the WTTP consent decree." I certainly think Smith was right about the consent decree constraints being outdated. In his opinion and mine, good-quality wildlife habitat, including waterfowl habitat, could be produced in far greater abundance by a restored stream than by any human-built developments on the mitigation land. As Smith pointed out, one of the primary purposes of the Mission Plan was to "enhance land for wildlife."[4]

It was ironic that the project had stalled because of an impasse between politicians who had professed support for natural resource restoration and a group of sportsmen—the Akers litigants—who were once considered heroes for halting the destructiveness of channelization. Dodd Galbreath did his best to resolve differences in a way that would lead to implementation of the Mission Plan, and in 1997 he was telling the press, "We've got all these pieces in place. We've just got to get through the gate."[5] But eventually such hopes evaporated like so much fog on a farm pond in September.

Myers's Dilemma

Caught in the middle of all this was Gary Myers, who had given our new principles of land stewardship in the northwest territories much-needed support on many other occasions but who sided with Akers on this issue. He seemed to think that somehow a reasonable solution could be found. At one point, he even thought—erroneously, as it turned out—that he had worked out a compromise with J. Clark Akers. As the *Dyersburg State Gazette* reported on February 20, 2000: "Myers said that if the Black Swamp alternative can be worked out, Akers has indicated that he will forgive the mitigation requirement needed to do the Stokes Creek project."[6] But Akers later said he did not recall making any promise to give up on the mitigation requirement. And I rather doubted that he ever did. He was not the sort to make such concessions.

The $12 million it would cost to meet Akers's demand was a sum the Corps of Engineers would have to agree to pay. But the corps would have no part of it, nor did it intend to purchase the remaining acreage since it saw no relationship between the original lawsuit and the intent of the Mission Plan.

The stubbornness of Akers not to agree to further work on the rivers without purchases for the mitigation lands, along with the corps's insistence that it had no money to do so, affected major work planned by the Basin Authority. In the fall of 1998, David Salyers, the new Basin Authority director, announced that the receipt of a $326,000 Environmental Protection Agency grant to restore 1.3 miles of Stokes Creek. The problem was that timber easements were required before Salyers's agency could begin work. In the agreement, the landowners had to agree never to convert the bottomland hardwoods to cropland (a condition already required by federal "Swamp Buster" laws).[7] The Basin Authority had easements for 215.76 acres, but these were not enough: the rest of the 416.54 acres were in the Tigrett WMA, and Gary Myers refused to allow these easements. Myers's decision really depended on what Akers wanted, and Akers said no.

Myers had tried his best, standing with us for ten years, but now he faced a new dilemma. He was losing credibility, standing to suffer the wrath of J. Clark Akers and his supporters if he did not side with them on the mitigation lands issue. Like everyone else, he wanted the state to have thirty-two thousand acres for the public's enjoyment. But in my view, neither he nor Akers should have used the Mission Plan as a hostage for the mitigation lands. Such wrangling might seem fair in some instances of "good ole southern politics," but in this particular case, there was too much at stake: vital principles of land stewardship and a geographic area that ultimately encompassed a million square miles and affected two million people. There was no doubt in my mind that whatever was decided with regard to the Obion–Forked Deer complex would apply to the rest of the rivers in West Tennessee and possibly beyond. Ironically, both Myers and Akers were members of the West Tennessee Tributaries Steering Committee, and both had ample acquaintance with the purpose of the Mission Plan and its importance to the state. The steering committee had cast affirmative votes for the Mission Plan, and this included channel projects on Stokes Creek and the Forked Deer River upstream at Tigrett.

Myers's allegiance to Akers was the root of his problem, in my opinion. He and the TWRA commissioners had already attempted to placate Akers by dedicating all the WMAs on the entire Obion River to him, whether or not these lay within his area of concern. Signs honoring Akers were placed at every WMA along the rivers. Certainly, Akers deserved generous praise for his earlier challenges against the WTTP, but his unyielding stance—unlikely to change no matter how many signs in his honor were erected—threatened the longevity of the rivers and all of the benefits that came with sustaining them in a responsible manner.

Siding with Akers and stopping the Mission Plan must have been an excruciating decision for Myers, one that he must have agonized over for a long while. He had to have known that the Mission Plan promised a far more correct and ethical approach to the crisis of the rivers than the one he ended up choosing. My own take is that the director thought that something could be worked out and that the state could ultimately have both the land and the plan. Indeed, this was what he indicated to me on numerous occasions. "We don't have the money now to do this project," he would tell me. Until that appeared possible, he believed, we had to continue to try changing minds and coming to agreements. But knowing the opposition, I cannot say that I shared his optimism.

Myers had stood with us, his field agents, too long not to feel remorse about his decision. He had even been urged to dismiss me from the agency, but he had resisted such pressure. I can only imagine his agony, since he had encouraged us for so long to "do the right thing."

A defining moment in this fray had finally been reached. Now the rivers and all that depended on them were held hostage.

Chapter 20

Conclusions

Where are we after more than thirty years of trying to "fix" Reelfoot Lake and restore the rivers? Not far—only a few yards in a mile-high climb. As of this writing, Reelfoot is still waiting for the preferred alternatives in the USFWS environmental impact statement to be implemented; plans to manage the rivers are still in gridlock.

Why did we fail? There are several reasons, but a particularly important one seems to be that the public's interest was not sufficiently engaged when it could have mattered most. We were unable to educate the citizens of Tennessee, so they did not clearly understand the issues. Consequently, they were unable either to express their full opinions or to provide the necessary support for protecting their interests in rivers and wetlands.

There was another factor that held us back: a dysfunctional government process. The very laws that were established to protect natural resources and the environment were used against them. In each case, one of the nation's most powerful environmental laws—the National Environmental Policy Act—was ultimately the mechanism used in the legal process to stall the conservation, management, and preservation of these natural resources. Something was seriously wrong with the system, and it needed to be fixed. Individual citizens cannot be held solely accountable for their lack of knowledge and confidence to voice their opinions. That responsibility lay with the gatekeepers: the state and federal natural resource departments and agencies. Without education from these capable professionals, many people will remain ignorant about rivers and wetlands. In this regard, there has been much to do.

The status of rivers affects every natural resource agency in the nation, but without a joint effort, little headway is likely to be made. Politicians, administrators, and the public must first be educated. Programs about the enjoyment of the great outdoors are not enough, although these should continue. Educational programs about rivers and wetlands should be a line item in the public relation budgets of every conservation organization and natural resource agency. A mass-media campaign needs to be not only audience-friendly but also deeply serious in the way it addresses this crisis. It is not an unrealistic task. This educational effort requires a specific goal, however, and that is to change minds and provide a means for the public to express how they feel about the management of waterways. The heroic, large-scale practices of shaping the land—damming streams, digging new channels, and remodeling the rivers and the surrounding landscapes—must be abandoned as abject failures. We simply cannot plunder nature in order to save it. Our footprints on the land must be better planned. All of our research in the past thirty years indicates that returning rivers to their natural state, wherever possible, is the best strategy for ensuring their long-term health. It has been a struggle for some of the most open-minded, forward-thinking wetlands managers to come to grips with this, break away from tradition, and confidently support the new thinking. But they eventually find that it is sound, and all of our education efforts should be geared toward understanding its importance and bringing everyone else along.

The role of elected officials in this process, of course, is absolutely critical. Still another factor thwarting river restoration was the failure of politicians to exercise proper leadership. Even when the agencies are of one mind and even when the people are well informed, elected officials in the state and federal governments cannot continue to avoid the issues and "pass the buck." They must actively confront the problems and forcefully carry out the public's will.

J. Clark Akers was, perhaps, legally correct in his opinion that the corps was obligated to purchase the rest of the mitigation lands. If he was right, Congress should have authorized the funding and put the corps back to work on Ben Smith's river restoration plan. Even Akers suggested as much at one point. In a March 1994 letter to Smith, Akers stated, "I whole heartedly support the consensus that the problems of the Obion and Forked Deer Rivers will not be cured by channelization or by piecemeal approaches which do not encompass the entire drainage area. The acquisition of the mitigation lands along the North and Middle forks of the Forked may speed up the construction of this test project by several years."[1] Personally, I have my doubts about Akers's sincerity and the concern for the rivers he expressed in that letter. I suspect that as long as he got his full thirty-two thousand acres of mitigation land, he did not really care about what happened to the rest of the watershed and to the other rivers of West Tennessee. Nevertheless, even if he was being disingenuous, Akers made a good point, particularly if the acquisition included enough land to protect the entire floodplain corridor—and if the management strategies were changed to mutually benefit the river and the citizens of the state.

But let us remember that Congress has not authorized the purchase of these lands, and therein lies another irony. If Congress could justify channelization—the epitome of poor river management—why on earth can it not authorize enough land to manage the river properly?

An alternative is to ask Congress to de-authorize the West Tennessee Tributaries Project and to authorize a new project in its stead, one with the cooperative engagement involving the reformulation of the WTTP and the Mission Plan. De-authorizing the WTTP would abolish all of the court provisions now plaguing the people of West Tennessee, including the consent order and the accompanying agreed order. Clearing this legal snarl would at least make it possible to start over with a cleaner slate. "It's an excellent thought," Bob Bryant, the president of the Quail Unlimited chapter in Jackson said at one his group's meetings during the summer of 2005. "But you know what happens when it reaches Nashville . . ." Dr. Bill Luckman, who stood nearby, agreed and hoped it would include enough wild lands for wide river corridors whenever it was done. "These are the lands we need if we plan to get the most benefits from rivers," he said.

While we are still at the beginning stages of solving a few of these intolerable problems, some piecemeal gains have been made. At Reelfoot Lake, additions to the buffer zone around the lake have incidentally provided us with an effective waterfowl refuge and more wildlife land. And we have salvaged a tentative, but effective, interim water-level management plan.

Otherwise, little has been accomplished to affect the ecology of Reelfoot Lake. At this juncture, the necessary construction of the new spillway is still being debated. The sparring involving politicians at the state and national levels, and Kentucky farmers at the local level, has not been benign. The combatants are still active, with those on one side determined to either stop the Reelfoot restoration project or have it their way. The U.S. Army Corps of Engineers is threatening to require yet another environmental impact statement related to the spillway design plans. This makes no sense. The EIS process is expensive and time-consuming, and neither time nor money remains to be squandered. If another EIS is done, it will be the third one undertaken, and the first should have been sufficient. Who knows how much further this folly will go before the public is served? One thing is sure: the State of Tennessee must make the necessary sacrifices and take a strong lead if what is left of Reelfoot Lake—or the rivers—is to be salvaged.

The Hatchie is one river that is readily salvageable; every effort should be made to restore it as nearly to its native characteristics as is reasonable and possible. It is the only

Waterfowl in a field at the edge of the floodplain away from the river channel. (Photo by Jim Johnson.)

river we have that points to the future possibilities for the remaining rivers. It is not too late to get ahead of the curve and salvage this river at the most favorable cost-benefit ratio. Every decade of waiting is a poor investment for the state. Even now, how long would it take a duck hunter, a conservation teacher, a visitor to the state, or even an accountant to weigh the options and make the right decision? Not long.

State and national agencies and organizations, both public and private, should join together in a coalition to select the critical priorities and work on the solutions for the problems of managing and sustaining these vital resources. Despite the froth of politics and legal maneuvers, and despite complaints in some quarters that nothing can be done, there are reasons to be optimistic. Congress, for example, has redefined the U.S. Army Corps of Engineers' mission for managing rivers and wetlands within its jurisdiction: the corps now has the authority to enhance and restore wetlands affected by their projects. Even the corps, once the mightiest builder of dams and levees, is beginning to adopt management practices that are more in line with what nature requires. It is for this reason that the corps could have been a progressive and effective partner in the Mission Plan.

Other agencies are moving ahead to meet the challenge. The U.S. Fish and Wildlife Service's Partners for Waterfowl and the USDA Wetland Reserve Program has preserved thousands of additional acres of wetlands along the river floodplains. Thousands of acres of floodplain wetlands have been acquired and protected during the same period by state and federal authorities. These lands act as buffers along rivers, and these help reduce the conflicts familiar to us. The Environmental Protection Agency's Section 319 non-point pollution program is in place to slow watershed pollutants from runoff. The state's 301D program provides funds to restore deteriorated small streams. The Conservation Reserve Program now holds the watershed soils on thousands of acres of farmland highly prone to erosion. All serve to decrease sediments, especially the heavy sands, from entering the floodplain. The West Tennessee Basin Authority has changed its river maintenance methods in favor of floodplain restoration, and the Mission Plan still awaits implementation.

Perhaps most important, the public is changing its mind about the best management practices. Duck hunters are not always thinking about what the next hunting season will bring; sometimes they think about how to assure hunting seasons for their grandchildren. Wetland management programs now in place have the support of people such as the farmers at Stokes Creek, the Dyer County sportsmen and other duck hunters, local ornithology chapters, and many who have longed to experience autumn hikes with their families along riverbanks. Individually, these supporters might have been mere voices in the proverbial wilderness, but united their voice can be formidable and give direction to how the state's river resources are managed. They should remind themselves that lawmakers need support from their constituents—people like them.

We should bear in mind that rivers represent us, the land users and the land stewards. We should not forget that the rivers of West Tennessee are the remnants of the last wild lands, the last vestiges of the region's native integrity and character. They signify and represent our strength and wealth. Within a very few years, there could be no evidence that these resources ever existed in West Tennessee—unless something is done, and soon.

One of the most important needs is for the governor's office to stay engaged so that the immense socioeconomic benefits that come with healthy rivers can finally be realized. It is time to move more aggressively, Ben Smith has argued: "It might require Governor [Phil] Bredesen [and his successors] to give the Mission Plan more high-level support [than previous administrations]; it might require the State to go into Federal Court and ask, based on today's improved conditions, that the chokehold of the old Consent Decree be broken."[2]

During the development of the *Lost Rivers* video, we talked to numerous citizens about the management of West Tennessee rivers. "It's crucial," said Martha Lyle Reed, who spent most of her youth along the Hatchie River, "that we change our attitudes [about the way we use the river]. We've used it only for our own interest. We've been shortsighted for future generations. We must live in harmony and respect for the rivers." Walter Powell, owner of the Powell Lumber Company, echoed her warning: "Landowners, timber owners, farmers, and everybody else are going to have to educate themselves, to take heart, and think about their fellow man—not to think about their own individual pockets—and work together."[3]

We should take heed of those who have spoken up for river wetlands. Aldo Leopold, Ernest Swift, and other great land-steward conservationists cautioned us more than half a century ago that a crisis was at our door.[4] Now, their warnings have come home to roost. The day to take stock and take heed is overdue. It is time for recompense, time to rethink what those with wisdom have said. It is time to "do the right thing."

Chronology

1796   Tennessee becomes state.

1811–12   New Madrid Earthquake creates Reelfoot Lake.

1818   Issac Shelby, ex-governor of Kentucky, and Gen. Andrew Jackson negotiate treaty with Chickasaws for purchase of West Tennessee lands.

         Barney Mitchell establishes regular trading routes on Forked Deer River to town of Jackson.

1820   First settlements established in Dyer County at Key Corner.

1823   Forked Deer River rerouted to Obion River by channel bypassing some twenty-five to thirty miles on river's lower reaches.

1826   Davy Crockett takes flatboats loaded with staves down Obion River en route to New Orleans.

         Steamboat *Red Rover* travels up Hatchie River to town of Bolivar.

1866   Andrew Atkinson Humphreys becomes first head of U.S. Army Corps of Engineers and conducts intensive projects to deepen Mississippi River for heavy barge traffic.

1883   North Carolina Land Grants open way for homesteading in extreme West Tennessee, including Reelfoot Lake.

1915   H. M. Golden surveys Reelfoot Lake public boundaries.

1916   Local watershed districts dig main channel of Forked Deer River through Dyer County.

         Private channelization on West Tennessee rivers begins.

1925   State establishes Reelfoot Lake Commission to permanently set boundaries of state land on Reelfoot Lake.

1931   Reelfoot spillway constructed as permanent structure.

1941   Lease Agreement for cooperative management of Reelfoot Lake between U.S. Fish and Wildlife Service and state.

1942   State and U.S. Fish and Wildlife enter seventy-five-year lease agreement to establish Reelfoot National Wildlife Refuge.

1948	Congress passes Flood Control Act, clearing way for West Tennessee Tributaries Project (WTTP).
1958–60	Carl Yelverton, state biologist, drafts waterfowl management plans for proposed site of Tigrett Wildlife Management Area (WMA).
	Major die-offs of bottomland hardwoods recorded along Forked Deer River at Tigrett WMA.
	Obion and Forked Deer rivers still boast excellent waterfowl hunting.
	Soybeans become accelerated commodity on world markets. Land clearing along river bottoms increased for soybean production.
1959	Congress passes Fish and Wildlife Coordination Act, requiring federal projects to consider and mitigate losses of wildlife lands.
1960	Tennessee Game and Fish Commission develops Gooch WMA on Obion River and purchases first tracts of Tigrett WMA on Forked Deer River.
	Corps of Engineers promises that WTTP will cause 95 percent reduction in acreage of freshwater swamps, lakes, and sloughs.
1961	Corps of Engineers begins work on Obion–Forked Deer WTTP.
1969	Congress passes National Environmental Policy Act (NEPA), requiring environmental evaluation wherever federal projects have major impact on environment.
1970	J. Clark Akers III and others file lawsuit to challenge corps's compliance with NEAP and other federal laws in implementation of WTTP.
	Thirty-two percent of WTTP completed.
1971	Three-quarters of Obion–Forked Deer floodplains devoted to agricultural development, with half of that in soybean production.
	U.S. Army Corps of Engineers, Memphis District, files first Environmental Impact Statement (EIS) in response to *Akers* lawsuit.
	Calvin Barstow, state waterfowl biologist, releases damaging report condemning WTTP as primary cause behind loss of bottomland hardwoods along Obion River.
1972	State legislature creates independent Obion River Basin Authority, directed by Richard Swaim, to maintain corps's WTTP on Obion–Forked Deer.
1973	*Akers* lawsuit upheld and injunction to halt WTTP issued.
	Experimental winter drawdown initiated at Reelfoot Lake.
1974	Tennessee Game and Fish Commission reorganized to become Tennessee Wildlife Resources Agency (TWRA).
1975	Corps files its second EIS in response to *Akers* lawsuit.
1978	Federal court declares corps's EIS answering *Akers* lawsuit inadequate.
1980s	Approximately 23 percent of 431,000 acres of bottomland hardwoods in West Tennessee survive land clearing.
1983	Frank Zerfoss submits TWRA's West Tennessee Tributaries Wildlife Mitigation Plan as required by Akers lawsuit.

1984	Tennessee Department of Conservation denies water quality permit required by corps to complete a portion of its work on South Fork Obion River.
	Frank Zerfoss and author present scathing report to TWRA director Gary Myer's staff condemning agency's policies for developing waterfowl units on Obion–Forked Deer river and development methods designated in WTTP mitigation plan. Report also advocates floodplain restoration.
	Author replaces Frank Zerfoss as supervisor of Region I northwest district.
1985	As result of *Akers* case and consent order, court requires corps, in part, to acquire 32,000 acres of land to mitigate impacts to water quality from WTTP channelization work.
	Injunction against WTTP temporarily lifted; corps budgets for restarting work on Forked Deer.
	Agreed order in *Akers* case outlines guidelines for Obion–Forked Deer Basin Authority work on Obion–Forked Deer.
	TWRA attempts extreme drawdown for Reelfoot Lake in May.
	Lawsuit filed in June against Reelfoot drawdown. Judge Odell Horton in Memphis District Federal Court orders injunction against drawdown and requires U.S. Fish and Wildlife Service to produce EIS.
1986	Corps attempts to restart WTTP dredging on South Fork of Forked Deer but cannot secure landowner easements, temporarily shutting down project.
	TDEC commissioner James F. Word yields to political pressure and asks corps to retract WTTP water quality permit it had previously denied in 1985. His request promptly withdrawn at insistence of landowners and conservationists.
	WTTP shut down for second time.
1988	Extreme discontent against corps's WTTP channelization of Obion–Forked Deer rivers develops. Governor McWherter's office and conservationists want to use more sensitive methods (stream-obstruction removal guidelines) but corps and channelization supporters refuse to change methods.
	Reelfoot Lake fifty-year management plan completed.
1989	U.S. Fish and Wildlife Service completes Reelfoot Lake EIS for managing water levels.
	Corps withdraws a water-quality permit application that would have restarted part of the WTTP.
1990	Last draft of Tigrett Floodplain Restoration Plan completed.
1991	Chester McConnell and author meet at Tigrett WMA and Ghost River to resolve their differences over artificial swamp issue.
	Tigrett Floodplain Restoration Plan released to public.
	Lost Rivers video produced to describe restoration of Tigrett WMA.
1993	Governor Ned McWherter names members of West Tennessee Tributaries Steering Committee under leadership of Ben Smith. Committee's first meeting held in State Capital Building in Nashville.

At its fifth meeting, steering committee votes unanimously to adopt river restoration conceptual plan.

Last field trips conducted on Tigrett WMA supporting floodplain restoration.

Corps of Engineers enters Reelfoot Lake project to assist in development and implementation of Reelfoot Lake Water Level Management Plan.

Corps completes Reelfoot Lake Reconnaissance Report for proposed management.

1994   Ben Smith and other Steering Committee members present Mission Plan for Reformulation of the West Tennessee Tributaries Project to Memphis district engineer.

Ben Smith presents Mission Plan to General Assembly as HJR-428.

1995   Corps completes Reelfoot Lake Feasibility Study.

Gov. Don Sundquist takes office in January and soon endorses Mission Plan. Meetings of Steering Committee and Technical Committee continue with new chairperson, Dodd Galbreath.

Corps announces that Mission Plan, including plan to restore Stokes Creek, feasible and eligible for federal funding.

Plaintiffs of 1972 WTTP lawsuit begin to apply Consent Decree conditions to Mission Plan demonstration projects.

Mission Plan placed on hold.

Meeting to "kill" the Tigrett Plan held at TWRA's office in Nashville.

1996   State Legislature abolishes Obion–Forked Basin Authority and creates West Tennessee River Basin Authority. Agency placed under administration of TDEC, with new mission statement that pledges "to restore, where practical, natural stream and floodplain dynamics," instead of maintaining work accomplished by WTTP. David Salyers replaces Richard Swaim as executive director of Basin Authority.

Last meeting on Black Swamp issue held in Nashville with J. Clark Akers, TWRA director Myers, and others.

1999   Congress passes Water Resources Development Act allowing corps to restore wetlands affected by their projects; Section 101 (b) allows construction of corps project at Reelfoot Lake.

Corps completes additional EIS for Reelfoot Lake following U.S. Fish and Wildlife Service's EIS (1989).

2005   WTT Mission Plan, Reelfoot Lake Water Level Management Plan, and Tigrett Floodplain Restoration Plan remain unimplemented.

In report requested by governor's office, Leigh Fredrickson and Sammy King condemn proposal to develop Black Swamp.

TDEC asks TWRA withdraw water quality permit to develop Black Swamp. TWRA does so.

Notes

Introduction

1. Sandra Postel and Brian Richter, *Rivers for Life: Managing Water for People and Nature* (Washington, D.C.: Island Press, 2003), 25.

1. The Five Rivers along the Mississippi

1. Katherine Carter Ewel and Howard T. Odum, eds., *Cypress Swamps* (Gainesville: Univ. Press of Florida, 1984), 372.
2. *State ex rel. Charles T. Cates et al. v. West Tennessee Land Company et al.,* April term 1913, Tennessee State Supreme Court Rulings, Reelfoot Lake Case, 614.
3. "West Tennessee Tributaries Project: Final Environmental Impact Statement," Department of the Army, Corps of Engineers, Memphis District, 1982, sec. 3, p. 57.
4. Ibid., sec. 3, pp. 75–95.
5. Spence Dupree, *Across the River* (N.p.: Self-published, 2001), 47.
6. "Interim Report: Obion–Forked Deer River Basin Survey," U.S. Department of Agriculture, 1972, 2–3. This report was requested by the office of Governor Winfield Dunn and cosponsored by the Tennessee Association of Conservation Districts 8 and 9. Copy in author's files.
7. Ibid., 21.
8. Earl Willoughby, "Citizen King: Cut-offs and Crockett River Days," *Dyersburg State Gazette,* Dec. 17, 1995.

2. River Users

1. I have drawn these numbers from various information sheets and surveys in my possession, including Tennessee duck hunting data compiled in 2004 by Don Orr (now retired from the U.S. Fish and Wildlife Service) under the auspices of the USFWS and the University of Memphis. Another key source is the Waterfowl Harvest and Population Survey data compiled by David Fronczak of the USFWS's Division of Migratory Bird Management in Columbia, Missouri.
2. "West Tennessee Tributaries Project: Final Environmental Impact Statement," 201.

3. This survey included a total of 7, 230,834 ducks; within that total, the number recorded for Tennessee was 465,408. "Midwinter Waterfowl Survey Mississippi Flyway," comp. Ken Gamble, U.S. Fish and Wildlife Service, Columbia, MO, 2002.

4. Quoted in Joseph P. Linduska, ed., *Waterfowl Tomorrow* (Washington, DC: U.S. Department of the Interior, Fish and Wildlife Service, 1964), 186.

5. Linduska, *Waterfowl Tomorrow*, 185.

6. According to a 1981 estimate, one of the last undertaken for the Obion–Forked Deer system, some 6,708 hunters made 74,890 trips that year to hunt ducks in those river bottoms. These numbers were cited by biologist Dan Sherry in a fourteen-page West Tennessee tributaries evaluation submitted to TWRA director Gary Myers on May 20, 1988. Copy in author's files.

7. Dupree, *Across the River*, 101.

8. Ibid., 96–97.

9. Ibid., 99.

10. Ibid., 106.

11. Timothy Broadbent, "Preliminary Aquatic Samples Forked Deer River," field notes, Tennessee Wildlife Resources Agency, Jackson, TN, 1993. The artificial swamp in this study was within the Tigrett WMA on the North Fork of the Forked Deer River. One of the two river meanders sampled was also at Tigrett, while the other meander was found on the river's South Fork.

12. The last survey by TWRA, conducted in 1981, estimated that 7,114 anglers made 164,278 fishing trips within the Forked Deer–Obion complex that year (cited by Sherry, report to Myers, May 20, 1988). Considering the size and length of the rivers, those numbers are actually quite low—perhaps a tenth of what they should have been.

13. "Interim Report: Obion–Forked Deer River," 2. I have been unable to obtain more recent data on flood damage, though such information probably exists in scattered form. It just happened that 1971 saw a flurry of interest in river issues by the Obion–Forked Deer Basin Authority—hence the report that yielded these statistics.

14. "Comprehensive Development Plan," Obion–Forked Deer Basin Authority, Jackson, TN, 1983, 10. Copy in author's files.

15. George Smith and M. B. Bandenhop, *An Evaluation of Environmental Quality: Opportunity Costs of Channelization and Land Use Changes in the Floodplain of the Obion–Forked Deer River Basin of West Tennessee*, Bulletin 552 (Knoxville: Univ. of Tennessee, Agriculture Experiment Station, 1975), 12.

16. Bruce A. Tschantz, Timothy R. Gangware, et al., *Private Levee Study in Tennessee: Extent, Hydraulic Impact and Management of Levees*, Research Project Technical Completion Report No. 114 (Knoxville: Univ. of Tennessee, Water Resources Research Center, 1987), 67–69.

17. Ibid., 100–101.

18. L. A. Weaver, and W. E. Hammitt, "Reelfoot Lake Visitor Use Study" (graduate research contract for the Tennessee Wildlife Resources Agency), Univ. of Tennessee, Dept. of Forestry, Wildlife, and Fisheries, 1986. Copy in author's files.

3. The Dynamics of a River and the Coming of Civilization

1. Juanita Clifton, *Reelfoot and the New Madrid Quake* (Asheville, NC: Victor Publishing Company, 1980), 80.

2. Blanche G. Peacock, "Reelfoot State Park," *Tennessee Historical Quarterly* 32, no. 3 (Fall 1973): 206–7, 209.

3. *Goodspeed's History of Tennessee: Dyer, Gibson, Lake Obion and Weakley Counties* (Nashville: Goodspeed Publishing Co., 1887), 844.

4. Earl Willoughby, "Citizen King, Cut-offs, & Crockett river days," *Dyersburg State Gazette,* Dec. 17, 1995.

5. See "The Hatchie" Web site—http://www.hatchie.com/—for information on Hatchie history.

6. David Crockett, *A Narrative of the Life of David Crockett of the State of Tennessee,* Tennesseana Editions (1834; reprinted, with introduction and annotations by James A. Shackford, Knoxville: Univ. of Tennessee Press, 1973), 147.

7. Ibid., 150–51.

8. Ibid., 174–75.

9. Ibid., 195.

10. Ibid., 196–98.

11. Ibid., 154.

12. J. T. McGill and W. W. Graig, "The Ownership of Reelfoot Lake," *Journal of the Tennessee Academy of Science* 8 (1933): 1, 14.

13. Crockett, *Narrative of the Life,* 194, 155.

14. Ibid., 181.

15. Ibid., 187–90.

16. Constance Rourke, *Davy Crockett* (New York: Harcourt, Brace and Company, 1934), 86–87.

17. Crockett, *Narrative of the Life,* 194.

18. McGill and Graig, "The Ownership of Reelfoot Lake," 13–15.

19. R. E. Lee Eagle, *Reelfoot Lake Fishing and Duck Shooting* (Nashville: McQuiddy Printing Co., 1915), 22–25.

20. Ibid., 10.

4. The Genesis of the Rivers and the Beginning of Channelization

1. Quoted in Jim W. Johnson and James L. Byford, *Lost Rivers* (Jackson, TN: D'Lomar Productions, 1992), video produced for Tennessee Wildlife Resources Agency.

2. "People and Events: Andrew Atkinson Humphreys," *American Experience* (PBS) Web site: http://www.pbs.org/wgbh/amex/eads/peopleevents/p_humphreys.html.

3. My account here of the corps's work on the Mississippi is indebted to John M. Barry, *The Rising Tide: The Great Mississippi Flood of 1927 and How It Changed America* (New York: Simon and Schuster, 1997).

4. Timothy Diehl, conversation with author during investigation on Tigrett Wildlife Management Area, Tennessee, Aug. 6, 1991.

5. The Wetland Managers and Their Work

1. Scott Yaich, "Ducks Out of Water," *Ducks Unlimited,* July/Aug. 2004, 68.

2. "Tennessee Conservation Strategy," 3rd ed., The Governor's Interagency Wetlands Committee and Technical Working Group, Environmental Policy Office, Tennessee Department of Environment and Conservation, Nashville, 1998, 22. Copy in author's files.

6. The Obion–Forked Deer Wildlife Management Areas

1. Carl S. Yelverton, "A Fish and Wildlife Conservation Plan for the Harrison Tract," report on Tigrett Wildlife Management Area, Tennessee Game and Fish Commission, Nashville, 1966, 4. Copy in author's files.
2. Quoted in Johnson and Byford, *Lost Rivers* video.

7. The West Tennessee Tributaries Project

1. Michael Hudoba, "Report from Washington," *Sports Afield,* Apr. 1951.
2. U. S. Army Corps of Engineers, "West Tennessee Tributaries Project: Final Supplement to the Final Environmental Impact Statement," Dept. of the Army, Corps of Engineers, Memphis District, 1982, plate 2-1. Copy in author's files.
3. Ibid., plate 1-20.
4. "Comprehensive Development Plan Obion–Forked Deer Basin Summary Report," Obion–Forked Deer River Basin Authority, Jackson, TN 1983, 3–7. Copy in author's files.
5. Calvin J. Barstow, "Fish and Wildlife Resources: Obion and Forked Deer Basin, Tennessee," preliminary report, Tennessee Game and Fish Commission, Nashville, 1970.
6. Calvin J. Barstow, "Impact of Channelization on Wetland Habitat on the Obion–Forked Deer Basin, Tennessee," in *Transactions of the Thirty-Sixth North American Wildlife and Natural Resources Conference, Wildlife Management Institute* (Washington, DC: National Wildlife Federation, 1971), 362–76.
7. Ernest F. Swift, *A Conservation Saga* (Washington, DC: National Wildlife Federation, 1967), 103.
8. *J. Clark Akers and William W. Dillon III v. The United States Army, et al.,* lawsuit filed in U.S. District Court, Jackson, TN, Apr. 23, 1970.
9. McConnell, who made this observation to me by telephone in August 2005, had fought the project in part by distributing flyers that resembled paper money. Bearing the image of a pig's head, they proclaimed: "Obion–Forked Deer River Channelization, One Hundred Million Dollar Ditch, Nation's Biggest Pork Barrel Project . . . Legal Pork Tender for Water Projects."
10. *Akers and Dillon v. Army,* 8.
11. Forest Durand to District Engineer, Memphis District, U.S. Army Corps of Engineers; this letter was an attachment to the final supplement of the corps's final EIS, submitted in December 1982.
12. Michael Mansur, "Draining Basins and Billfolds," *Memphis Commercial Appeal,* Mar. 4, 1984.
13. McConnell to author, e-mail, Mar. 2006.
14. Ibid.
15. From a summary of the West Tennessee Tributaries Project in "The Mississippi River Commission Meeting at Vicksburg, MS," U.S. Army Corps of Engineers, Apr. 3, 2001, 407. Copy in author's files.
16. "Comprehensive Development Plan," 23, 32.
17. "An Overview of the Activities and Responsibilities of the West Tennessee River Basin Authority," report by Basin Authority, Department of Environment and Conservation, State of Tennessee, Humbolt, TN, 2003, 2. Copy in author's files.

18. Ibid.
19. Political support for empowering the Basin Authority to meet these objectives seemed to be present in 1987, according to an extensive University of Tennessee study on private levees. See Tschantz, Gangware, et al., *Private Levee Study in Tennessee,* 151.

8. Reelfoot Lake

1. Lexie Leonard, *Reelfoot Lake Treasures* (Tiptonville, TN: The Lake County Banner, 1991), 2.
2. McGill and Craig, "The Ownership of Reelfoot Lake," 15.
3. *Cates v. West Tennessee Land Company,* 650.
4. Ibid., 582.
5. Ibid., 582–83.
6. Johnson, Brown, et al., "Reelfoot Lake Fifty Year Management Plan," 7.
7. Wintfred E. Smith and T. D. Pitts, "Reelfoot Lake: A Summary Report," Univ. of Tennessee at Martin, 1982, 26. Copy in author's files.
8. For a full account of these events—the causative factors, the violence, and the aftermath— see Paul J. Vanderwood, *Night Riders of Reelfoot Lake* (1969; reprinted, with a new afterword, Tuscaloosa: Univ. of Alabama Press, 2003), esp. 8–74.
9. Vanderwood, *Night Riders,* 97–148; Smith and Pitts, "Reelfoot Lake," 26–27.
10. "Final Report of the Reelfoot Lake Commission to Henry H. Horton, Governor, State of Tennessee," Paris, TN, Aug. 10, 1931.
11. Smith and Pitts, "Reelfoot Lake," 27.
12. "Final Report of the Reelfoot Lake Commission." 42.
13. Virtually all accounts of the Reelfoot Lake area as it existed in the nineteenth century mention regular flooding from the Mississippi River. According to Smith and Pitts ("Reelfoot Lake," 29–30), private levees had been built along the Mississippi "since well before the Civil War, but they did not form an integrated system." Coordinated levee construction would come after the formation of the Mississippi River Commission in 1879 and the establishment of levee districts. For all practical purposes, Smith and Pitts write, the "final separation" of Reelfoot from the Mississippi through levee construction was complete by 1920. However, they note, while these developments allowed agricultural interest to draw ever closer to the lake, the problem of flooding on agricultural land "did not end with the completion of the levee system." They add, "Solutions to such problems are difficult."

10. The Tigrett Floodplain Restoration Plan

1. Jim W. Johnson, "Tigrett Wildlife Management Area Floodplain Restoration Plan," report, Tennessee Wildlife Resources Agency, Nashville, 1990. Copy in author's files.
2. Neil A. Miller, "Effects of Permanent Flooding on Bottomland Hardwoods and Implications for Water Management in the Forked Deer River Floodplain," *Castanea* 55, no. 2 (June 1990): 111.
3. Quoted in Kathy Krone, "Up the Creek ..." *Dyersburg State Gazette,* Nov. 10, 1996.
4. Ibid.
5. Quoted in Johnson and Byford, *Lost Rivers* video.
6. One issue on which the Stokes Creek farmers and I agreed was the need to eliminate the artificial swamps. But Ron Fox told me that the local sportsmen would "eat me alive" if I

tried to drain the swamps. Later, on March 1, 1991, Ralph Gray and I met with the Dyer County sportsmen's conservation club—the West Tennessee Wildlife Resource Association—to disarm the odd rumors that we might destroy their hunting area. Pulling no punches, Ralph and I told them about the entire intent of the proposed Tigrett restoration and addressed all their questions. If there was anyone who objected to our plan, I never knew it. In fact, from that point on, we listed them as supporters of the Tigrett plan.

7. Quoted in Johnson and Byford, *Lost Rivers* video.

8. "Environmental Restoration Kissimmee River, Florida: A Final Integrated Feasibility Report and Environmental Impact Statement," Dept. of the Army, Corps of Engineers, Jacksonville District, 1991.

11. Rejection of the Tigrett Plan

1. Under President Bill Clinton, who took office in 1993, federal policy toward wetlands would incorporate the notion of quality as well as quantity.

2. Aldo Leopold, *A Sand County Almanac* (New York: Oxford Univ. Press, 1949), 224–25.

3. Chester McConnell, D. R. Parsons, et al., *Stream Obstruction Removal Guidelines* (Bethesda, MD: American Fisheries Society, 1983), brochure.

12. The Drawdown Strategy at Reelfoot Lake

1. Smith and Pitts, "Reelfoot Lake: A Summary Report."

2. G. M. Denton and W. R. Robbins, eds., "A Clean Lake Study: The Upper Buck Basin of Reelfoot Lake, Tennessee," Tennessee Dept. of Health and Environment, Division of Water Management, Nashville, 1984. Copy in author's files.

3. This video was produced around 1985 by TWRA's Information and Education Section, Nashville.

4. E. O. Gersbacker and E. M. Norton, "A Typical Plant Succession at Reelfoot Lake," *Journal of Tennessee Academy of Science* 4 (1939): 230–38.

5. Wesley Henson and Andrew Sliger, "Aquatic Macrophites of Reelfoot Lake," Department of Biology, University of Tennessee at Martin, 1983–86.

6. Johnson, Brown, et al., "Reelfoot Lake Fifty Year Management Plan," 63–73.

7. Interview, Gray's Camp, Reelfoot Lake, July 2005.

8. The Lake County Levee District was authorized by the Tennessee legislature at the instigation of James C. Harris, who had plotted to take over and drain Reelfoot Lake. See McGill and Craig, "The Ownership of Reelfoot Lake," 20.

9. "An Act to Regulate and Control the Waters of Reelfoot Lake," House Bill No. 722, Tennessee General Assembly, Mar. 29, 1929, 1.

13. Implementing the 1985 Extreme Lake Drawdown

1. Late in 2004, reporter John Brannon ("Pinion Said Lawsuits Could Spell End of Public Use of Reelfoot Lake," *Union City Daily Messenger,* Nov. 23, 2004) wrote about the still-unresolved lawsuits: "For Tennessee, the stakes are high in two Reelfoot Lake lawsuits pending in U.S. District Court in Jackson." Brannon quoted State Rep. Phillip Pinion as saying that an adverse ruling in the suits would be devastating. "As I understand it, if the plaintiffs prevail," Pinion told the reporter, "it would make Reelfoot Lake a privately-owned lake with the exception of about 4,000 acres the state would retain. It would effectively

take away the use of Reelfoot Lake as a public lake. I am also very concerned what effect it would have on a new spillway, especially since Congressman John Tanner, Gov. Phil Bredesen, state Sen. Roy Herron and myself have worked so hard to get funds to build it."

2. Greg Denton, "Sunset for Reelfoot Lake?" *Tennessee Conservationist,* July/Aug. 1984, 14.

3. These totals are contained in a letter from University of Tennessee professor John C. Rennie to Paul R. Brown, TWRA wildlife manager, Jan. 10, 1996. The letter was a belated response to questions raised by the TWRA to a report it had commissioned a decade earlier: L. A. Weaver and W. E. Hammitt, "Reelfoot Lake Visitor Use Study," Univ. of Tennessee–Knoxville, Department of Forestry, Wildlife, and Fisher graduate research contract for the Tennessee Wildlife Resources Agency, 1986. Copies of the letter and report are in the author's files.

14. Wrestling with the Fate of Reelfoot Lake

1. Cook has since become the district supervisor in charge of Reelfoot National Wildlife Refuge and two other refuges.

2. Johnson, Brown, et al., "Reelfoot Lake Fifty Year Management Plan," xi.

3. U.S. Fish and Wildlife Service, "Reelfoot Lake Water Level Management: Final Environmental Impact Statement," U.S. Department of the Interior, Washington, DC, 1989.

4. See John Brannon, "Bredesen, Others Looking Out for Reelfoot Lake," *Union City Daily Messenger,* Oct. 24, 2004.

5. In a cover document (signed and dated Dec. 23, 1999) that was attached to the corps's EIS on Reelfoot ("Reelfoot Lake, Tennessee and Kentucky: Final Feasibility Report and Environmental Statement," U.S. Army Corps of Engineers, Memphis District, 1999), Lieutenant General Joe N. Ballard, the corps's chief of engineers, went on record with a measured condemnation of his agency's own past practices and their effect on Reelfoot Lake: "Flood control and drainage improvements [mainly by the corps] in the [Mississippi River] basin have dramatically impacted the quality of fish and wildlife habitat. Construction of the Mississippi River levees in the 1930s stopped the almost annual recharge of the lake by overflow from the Mississippi River." He went on to note that construction of the original spillway at Reelfoot and stabilization of the lake's water levels had resulted in land clearing and agricultural conversion around the lake, which in turn "contributed to an unusually high rate of sediment deposition in the lake, which is reducing the value of the lake's aquatic habitat and the lake's value as a flood attenuation system."

6. Volume 1 of the corps's report, "Reelfoot Lake, Tennessee and Kentucky," was released in 1999, although three supplementary volumes (2–4) were actually released a year earlier. The delay in the release of the first volume was mainly because of the conflict with the Kentucky farmers and the subsequent squabble between Congressman John Tanner and Senator Mitch McConnell of Kentucky.

7. Randy Cook to author, telephone conversation, July 1, 2002.

8. Our objections drew on a report, "Reelfoot Lake Management: U.S. Corps Feasibility Study vs. Preferred Biological Management," that Cook and I had prepared for Myers in 2002 (copy in author's files). This five-point report compared the USFWS Reelfoot Lake EIS with the corps's revised plan for a three-foot lake drawdown rather than a four- to eight-foot drawdown. The corps called for a March 1 rather than an April 15 drawdown, limiting water-level fluctuation rather than instituting a dynamic water-level fluctuation,

and lowering peak water-level fluctuation of elevation to 283.2 feet above mean sea level, as opposed to the USFWS recommendation of 284. The corps's plan also raised the possibility that it would be in charge of the spillway operations. In general, the corps's proposals seemed to favor the interests of the Kentucky farmers over the health of the Reelfoot Lake ecosystem.

9. Brannon, "Pinion Said Lawsuits Could Spell End of Public Use of Reelfoot Lake."

15. Reviving the Dwindling Interest at Tigrett

1. Tony Campbell to Ron Fox, Mar. 18, 1992. Copy in author's files.
2. In addition to TDEC and TWRA, the U.S. Fish and Wildlife Service, the Tennessee Department of Water Pollution Control, and the Wildlife Management Institute were all represented on the committee.
3. Frank L. Zerfoss, "West Tennessee Tributaries Mitigation Lands Wildlife Management Plan," Tennessee Wildlife Resources Agency, Nashville, 1990. This document, a revision of a version that had been submitted seven years earlier, summarizes the options that were before the committee in the 1980s.
4. Report from Dan Sherry and the committee members to Gary Myers, May 20, 1988, 5. Copy in author's files.
5. Ibid., 3.
6. Ibid.
7. Ibid., 6.
8. Quoted in Johnson and Byford, *Lost Rivers* video.
9. Comments at a public meeting, Feb. 11, 1993.
10. See George F. Smith and M. B. Badenhop, *An Evaluation of Environmental Quality: Opportunity Costs of Channelization and Land Use Change in the Floodplain of the Obion–Forked Deer River Basin of West Tennessee,* Bulletin 552 (Knoxville: Univ. of Tennessee, Agricultural Experiment Station, 1975). This report was based on Smith's dissertation at UT, completed the previous year.
11. Postel and Richter, *Rivers for Life,* 12.
12. Ibid., 9–10.
13. James M. Oliver and Reza Pezeske, "Factors Influencing Survival and Productivity of Forested Wetlands in West Tennessee," research paper, Univ. of Memphis, 1995.
14. Ibid.
15. William J. Wolfe and Timothy H. Diehl, *Recent Sedimentation and Surface-water Flow Patterns on the Floodplain of the North Fork Forked Deer River, Dyer County, Tennessee,* Report 92-4082 (Nashville: U. S. Geological Survey, 1993).
16. Timothy Diehl, personal communication regarding the U.S. Geological Survey's 1993 investigations on Tigrett WMA. Diehl made these observations to me during a field trip with me at Tigrett, Aug. 6, 1993.

16. The Burial of the Rivers

1. Quoted in Kathy Krone, "Preserving the Forked Deer," *Dyersburg State Gazette,* Mar. 22, 1993.
2. Smith, Larry J., "The Story of Tigrett: A Swamp's Fights for Its Life," *Cumberland-Harpeth Audubon Society Newsletter* (Clarksville, TN), Nov. 1992.

3. David Shankman and L. J. Smith, "Stream Channelization and Swamp Formation in the U.S. Coastal Plain," *Physical Geography* 25, no. 1 (2004):. 22–38.

4. Quoted in Peter Schutt, "Controversy Surrounds Plan to 'Restore' WMA Swamp," *Mid South Hunting & Fishing News,* Jan. 29–Feb. 12, 1993, 12.

5. Peter Schutt, "Always See Both Sides," *Mid South Hunting & Fishing News,* Feb. 13–26, 1993, 2.

6. James A. Mallard, letter to the editor, *Dyersburg State Gazette,* Apr. 9, 1993.

7. Wilson lost his bid for the Republican nomination to Nashville surgeon Bill Frist, who went on to win in the general election in 1994.

17. Black Swamp

1. Leigh Fredrickson and Sammy King, "Field Evaluation of Black Swamp," Aug. 11, 2004, 11. While this document acknowledges the assistance of the Tennessee Department of Environment and Conservation in producing its evaluation, it is not clear from the copy in my possession which state agency actually sponsored the report.

2. Leigh Fredrickson, attachment to "Field Evaluation of Black Swamp," 5.

3. Fredrickson and King, "Field Evaluation of Black Swamp," 1.

4. Fredrickson, attachment to "Field Evaluation of Black Swamp," 7.

5. Devin Wells to Chester McConnell, Dec. 22, 2004. Copy in author's files.

6. Before the upheaval over the withdrawal of the 401 permit for the development of Black Swamp had a chance to subside, TDEC commissioner Betsy Child advised director Gary Myers that TDEC would issue a 401 permit for a similar but private project on the Tom Rice land immediately north of White Lake Refuge. Area manager Carl Wirwa and I recommended against issuing a permit for the Rice property, and all of the managers in my district argued against a permit for either of the projects because it would put TWRA in the precarious position of allowing levees on one property while denying permits for projects with similar environmental impacts. The Rice land development would not only end up killing the forest on that land but would also kill trees on White Lake Refuge, which was public land. We had already seen such results in places like the Gooch and Tigrett WMAs, as well as throughout the rivers. Director Myers called a meeting on January 12, 2005, with TDEC and the state Attorney General's Office to address this issue. If the state did the "right thing" and applied the decision made at Black Swamp, then it would be setting an ethical precedent that could be of tremendous benefit, both for the future of river resources and for the people of the state. Nevertheless, this had not happened by August 2005. In fact, it appears that the opposite will happen. If it does, it will jeopardize not only the reputation of two state conservation agencies but also go against all that we had tried to do during the last three decades, not to mention endangering the future restoration of West Tennessee rivers.

18. Ben Smith's West Tennessee Tributaries Mission Plan

1. Ben's presentation at the Nashville meeting, which Hurst and I attended as "observers," was also well received.

2. Tschantz, Gangware, et al., *Private Levee Study in Tennessee.*

3. "A Mission Plan for Reformulation of the West Tennessee Tributaries Project," Tennessee State Planning Office, Office of the Governor, 1994, 2.

4. Ibid.

5. When I asked Ben Smith how he had developed the concepts in the Mission Plan, he told me that he had drawn heavily on the Tigrett Floodplain Restoration Plan but had also used advice from the South Florida Water Conservation District based on its involvement in the Kissimmee River restoration. In addition, he had incorporated the *Stream Obstruction Removal Guidelines* that Chester McConnell and others had developed. Smith had covered all the bases. Ben Smith, in notes to and telephone conversations with the author, July–Aug. 2005.

6. Ben Smith, "Tennessee's Obion and Forked Deer Rivers: A New Mission for the U.S. Army Corps of Engineers," a presentation to the HJR-428 Study Committee, Jackson, TN, Dec. 21, 1994. A copy of Smith's handwritten speech, with his corrections, is in my possession.

7. Ben Smith to author, telephone conversation, Aug. 2005.

8. "Basin Accord: State Panel Repudiates River 'Channelization'" (editorial), *Memphis Commercial Appeal,* Apr. 15, 1994; "A McWherter Success" (editorial), *Nashville Tennessean,* Apr. 12, 1994.

19. The Rivers Held Hostage

1. "Tennessee Wetlands Conservation Strategy: Current Progress and Goals," 2nd ed., Tennessee Environmental Policy Office, Nashville, Tennessee, Jan. 1996, 110.

2. Kathy Krone, "Will Corps Re-engineering Clog Local River Project?" *Dyersburg State Gazette,* Apr. 20, 1995.

3. John Tanner to Milton H. Hamilton Jr., Mar. 11, 1998. Copy in author's files.

4. Ben Smith in notes to author, Oct. 7, 2004. Author's files.

5. Quoted in Kathy Krone, "Hamilton Says Local Support Key to WTTP," *Dyersburg State Gazette,* June 27, 1997.

6. Kathy Krone, "A 27-Year-Old Lawsuit Threatens to Stop a River Restoration Project in Dyer County," *Dyersburg State Gazette,* Feb. 20, 2000.

7. "Tennessee Gets EPA Grant for Stokes Creek Project," news release, Tennessee Department of Environment and Conservation, Sept. 21, 2001.

20. Conclusions

1. J. Clark Akers to Ben Smith, Mar. 8, 1994. Copy in author's files.

2. Ben Smith to author, typewritten note, summer 2005. Author's files.

3. Reed and Powell quoted in Johnson and Byford, *Lost Rivers.*

4. Leopold, *A Sand County Almanac;* Swift, *A Conservation Saga.*

Index

Matthews, Sen. Harlan, 199
Mayfield Creek, 191
McClinton, James, 42, 43
McConnell, Chester, 72, 75, **76**, 77, 170, **215**;
    Black Swamp permits, 193; Ghost River
    Field trip and, 124–26; stream obstruction
    removal guidelines and, 165
McConnell, U.S. Sen. Mitch, 161, 223n14
McKinney, Aubrey David, 119, 181
McWherter, Gov. Ned Ray, 117, 148, 193, **215**;
    Ben Smith and, 196; Black Swamp and
    Mission Plan and, 193; channelization and,
    199; Mission Plan and, 199; reformulation
    of WTTP and, 201; WTTP and, 197
Michael, Lewis, 136
Middle Fork Hunt Club, 73
Miller, Neil, 111, 164
minor drawdown, 130
Mississippi Flyway, 48, 54, 218n2; duck inven-
    tories and, 20; state waterfowl objectives
    for, 50; West Tennessee river bottoms and,
    20; wintering grounds and, 47; vanishing
    habitat and, 59
Mississippi River
—dynamics on Reelfoot: flooding and; dead and
    dying cypress and, 88; fish and aquatic life
    stocking and, 88
—early history of: catastrophic floods and, 27–31;
    Chickasaw treaty and, 35; extreme floods
    and, 28–29; great flood described on, 29;
    hunting and trapping along, 32; Reelfoot
    Lake, 86–90; relationship between rivers
    and people, 31–37; early inhabitants and,
    31–37; early settlements and, 32; oxbow
    formation and succession of, **29**; favorable
    for, 29; as last frontier and, 35–37; purchase
    and settlement of, 35; Reelfoot Lake and,
    36; pertinence of early history to, 36–37;
    river trade, 32–34; Barney Mitchell and;
    David Crockett and, 33–34; Red Rover and,
    33; towns and cities along, 36
—recent history xix, xx, **3**, **30**, 201, **213**, 221n8;
    agriculture production in, 46; dynamics of,
    88; Forked Deer River and, 11; General
    Andrew Atkinson Humphreys and, 44;
    levees along, 137; Loosahatchie and, 6;
    Mission Plan and, 201; Gary Myers land
    acquisition along, 107; Obion-Forked Deer
    and, 10; Obion River and, 14; oxbows and,
    86; Reelfoot Lake and, 92, 93, 94, 129;

restoration of, 171; subdue-and-control
    mindset and, 36, 45, 46; Tigertail and,
    192; Mississippi River levees, 45, 90, 94,
    137; stress and decline along, 43; "taming"
    the, 45
mitigation lands, 14, 164, 194, 208, 224n15;
    administrators and perceived threats to,
    119; artificial swamps and, 125; Akers
    and, 78; large "gift" of compromise and,
    74; terrible impasse and, 79; author's
    presentation and, 105; Black Swamp and,
    187–89; controversy and, 100; corps
    acquisition and, 78; authorization to
    purchase and, 97; litigants' position on,
    165; management design of, 75; objects to
    author's presentation and, 106; ownership
    and management of, 75; perceived threats
    to, 119; public control of river corridor
    and, 105; mitigation plan held hostage by
    constraints on, 203–5; significant meeting
    on, 166; TCL supports WTTP because of,
    181; Frank Zerfoss and, 103
"moist soil" plants and, 52, 53
mother of West Tennessee rivers, 27, 135
"Mr. Legend," 146, 147, 155, 156
Myers's dilemma: Akers's stubbornness and, 204;
    Black Swamp and, 204; concessions to J. Clark
    Akers and, 204; on "do the right thing" and,
    205; "good ole southern politics" and, 205;
    mitigation requirements and, 204; timber
    easements and, 204; mitigation land and, 204;
    root of problem and, 205; timber easements
    and, 204; on rivers held hostage, 205
Myers, director Gary, 145; author requests with-
    draw of Tigrett permit and, 179; author on
    Tigrett plan, 117; Black Swamp field meet-
    ing and, 187–90; Black Swamp Nashville
    meeting and, 191–93; Harvey Bray's replace-
    ment and, 101; Campbell meeting to "kill"
    Tigrett Plan and, 181–83; dilemma of, 203,
    204; *Lost Rivers* video and, 169; Nashville
    staff Tigrett plan meeting and, 119–20; new
    lands promised by, 173; public resistance
    and leadership and, 144; Reelfoot draw-
    down and, 148–50; Reelfoot drawdown
    and buffer zones and, 153; Reelfoot spillway
    issues and, 159–61; Stokes Creek farmers
    and, 114, 115, 179, 180; Frank Zerfoss and
    author's meeting with, 104–7; WTTP staff
    recommendations and, 164–66

Reelfoot Lake (cont'd)

lake administrators and, 131; local pre-
drawdown (early 1980s) and, 131; organic
sediments and, 136; oxygen deficits and,
134, 135; promises from, 134; sediments
and, 134–35; Reelfoot Task Force and;
"Reelfoot Management Report" and, 132;
Reelfoot Task Force and, 132

—impact of levees: animal and plant life and, 137;
biological communities and, 137; cypress
trees and, 137–38; early 1900s levee-building
and control of nature and, 137; ecological
equilibrium and, 137; food chains and, 137;
natural rise and fall of floodwaters etc. and,
137; wetland degradation and, 137; water
and nutrients and, 137

—Fifty-Year Management Plan: benefits and,
157; Paul Brown and author and, 157; goals
and objectives and, 156; implementation
requirements of, 157; Joint Senate Resolution 235 and, 157; "Mr. Legend" and,
157

—formation: as an earthquake lake, 86, 87; as an
oxbow and cypress swamp, 87; as a nutrient-
rich sink, 87

—local opposition: water levels and, 157–
58; buffer zones and, 157–58; political
opposition and, 158; Kentucky Department
of Conservation projects and, 158;
Kentucky farmers and, 158; Lake County
Commission and, 158; land acquisition and,
158; Kentucky buffer zone land and, 158

—managing natural wetland succession: biological
age of, 139; drawdown techniques and, 139;
environmental factors common to, 138;
examples of, 139; general rule of thumb for,
138; methods for, 139; relationship between
wetlands and native rivers and, 139; stages
of, 139

—river flooding; Frank Gooch comments and, 94;
flooding at Walnut Log and, 91–92; Ronnie
Johnson comments and, 94; lake managers
and, 94; Onice Strader and 1937 flood and,
92–93, 94; towns and cities along, 36; Reelfoot spillway project and, 159, 161; revised
water-level management plan meeting, 160;
Congressman John Tanner and, 160, 161;
TWRA and USFWS question authority of,
159–61; USFWS environmental impact
statement and, 158–60

—spillway: biological management failures and,
91; on conflicts and, 91; fish entering lake
and, 160; Kentucky farmers and, 209; lake
levels as boundaries and, 91; new design
and, 160; Reelfoot Lake Commission and,
91; Representative Phillip Pinion and, 161;
Congressman John Tanner and, 161; U.S.
Corps and construction of, 161, 209

—Tanner-Hamilton political team: lake property
owners and, 147; as lake supporters and,
147–48; "Mr. Legend" as symbolic leader
and, 148–49; as legislators, 147; local
leadership needs and, 148; local people
and, 149; as local politicians, 147; as
Obion County residents and, 147; public
confidence and, 148

Reelfoot Lake Treasures (Leonard), 83

"Reelfoot Management Report", 132

Reelfoot National Wildlife Refuge, 84, 91, 150,
155, **213**

Reelfoot ownership and Night Riders: court cases
and, 89–90; on draining the lake, 89; James
C. and Judge Harris and, 89; kidnapping
and murder and, 90; local militia and, 90;
Night Riders and, 90; Quentin Rankin and,
89–90; R. Z. Taylor and, 89–90

Reelfoot physical features: ecological changes
in, 85; biological communities of, 84–85;
major basins of, 84; vicinity map of, **84**

Reelfoot Task Force, 132

Reelfoot Technical Committee, 133, 158, 201, 202

Reelfoot waterfowl hunting: early statistics and
anecdotes, 88–89

Richter, Brian, 171

riparian rights: "Mr. Legend" and, 146–48, 161

river corridors: ecosystems of, 55; home ranges in,
55; island remnants of, 55; stewardship of, 55

river crisis: land and citizens affected by, xvii;
manmade ditches and, xvii; problems of,
xvii; river miles vanish from, xviii; to make
sense from, xviii; outlook dismal for, xviii;
wetland managers and, xviii

*Rivers for Life: Managing Water for People and
Nature* (Postal), 171

"River Rats," 14, 19, 106, 175

river users: abusers as, 18; artificial swamps and,
22, 23; bottomland hardwood and, 24;
comparative surveys of, 22, 23; court cases
and, 25; farmers and landowners, 23–25;
fishable waters and, 22; fishermen, 2, 23;

floodplain use by, 24–25; game hunters as, 19–22; hardwoods and game abundance for, 19; land clearing by, 24; levees and, 24–25; "levees wars" and, 25; Mississippi flyway and, 20; national resource banks for, 18; other users of the rivers, 25, 26; quality of, 23; response to; silent majority and, 18; soil conservation and, 24–25; sportsmen and, 17; statistics of, 26; surveys of, 23; Twin Rivers club and, 20–22; waterfowl hunters, estimates of, 19–20; veteran hunters at Reelfoot Lake as, **21**

Royal, George, 176

Rutherford, Henry, 29, 32

Salyers, David, 81, 204, **216**

Samburg, 32, **84**, 92, 130

Sasser, Sen Jim, 199

Scherer, Col. Jack, 160

Schutt, Peter, 184

"scoping" meeting, 144

sediment rates, 172

settlers, 30–36, 39, 88, 183; David Crockett account and, 33; during 1823, 33; footsteps of, 30; forest clearing and, 39; land unhealthy for, 88; Larry Smith artificial swamp arguments about, 183; raw earth and, 39

Shelby, Governor Isaac, 35, 213

Shelby Lake (the "Scatters"), 157

Sherry, Dan, 115, 120, 122, 163, 164, 181, 218n2, 224n15; with author on Tigrett plan, 120, 121, 185; as environmental biologist, 115; flight over Tigrett and, 114; Myers and, 115; meeting with Myers and author and, 117; Ben Smith and, 197

Serra Club, 173, 183, 184

Sliger, Andrew, 132, 134

Smith, Ben, 120, 196, 208, **215, 216**, 225n18, 226n18; author's first meeting with, 195; author's West Tennessee meeting with, 197; background of, 196; Gov. Alexander's Safe Growth Team and, 196; Gov. Phil Bredesen and, 211; Ron Fox and, 197; Dodd Galbreath and, 202; governor's office and; presentation by; 195; Julius Johnson and, 203; Kissimmee River EIS and, 197; Lost Rivers video and, 197; Mayfield Creek restoration and, 196; Gov. Ned Ray

McWherter and, 196; Gov. McWherter's request to reactivate WTTP and, 197; Mission Plan goals and objectives and, 198; news media and, 199; Gov. Sundquist and, 204; Tigrett Floodplain Restoration Plan and, 197; Bruce Tschant private levee report and, 197; WTT Steering Committee meeting and, 198; WTT Steering Committee's technical committee and, 197–98; WTT Steering committee, appointment of, 197; WTT Steering Committee approves, 200

Smith, George F., 171

Smith, Larry, 183

Smith, Wintfred, 123, 132, 150

South Florida Kissimmee River restoration project, 121

Sports Afield (Hudoba), 70

stagnant swamps, xvii, xviii, 5, 11, 23, 60, 102, 109, 111, 113; artificial channels and, 109; fishing and, 23; levees and, 102; perennial water and, 113; Tigrett swamps and, 135; Waterfowl habitat and, 11, 60; West Tennessee rivers and, xvii, xviii

Soil Conservation Service, 45, 69, 79, 127, 198; author's work group and, 80; Public Law 566 and, 7; small-stream channelization and, 77; soybean industry, 24, 43, 72; artificial swamps and, 124; bottomland hardwoods and, 71, 72; channelization and, 46; demand for, 24, 72; floodplain clearing and, 24, 43, 70, 71, 100; floodplain production and, 10, 23, 25, 47; levees and, 24; Stokes creek farmers' flooding problems and, 109

State Gazette, 11, 168, 169, 183, 202, 204

Steamboats, 32, 33, **213**

St. Francis Lake, 87

Stokes Creek, 109, 200, 210, **216**, 221n10; Basin Authority and, 204; Tony Campbell and, 181; channelization and, 114; farmers and, 116, 163, 173, 179, 182; Commissioner Milton Hamilton and, 203; Representative Harold Holt and, 114; lateral ditch and, 114, 125, 182; Mission Plan and, 205; Gary Myers and, 115, 179, 180, 204; Gov. Ned Ray McWherter and, 117; public meetings about, 116, 117, 186; Congressman John Tanner and, 203; Tigrett plan and, 113, 114, 116, 186; Tigrett restoration team and, 185, 186; Tigrett Technical Committee and, 122

Don Porter plan and, 185; Stokes Creek
issues and, 185; Chief Taylor meeting and,
185

Tigrett Technical Committee: formation of, 110;
members of, 123; purpose of, 123; Tigrett
plan and, 111, 120, 121, 122, 163

Tigrett Technical Committee on artificial
swamps: author's solutions at Annie Laurel
James farm and, 125; author's views on,
124–25; critical issue and status of, 124;
compromising principles of, 124–25; cypress
swamps and, 127; Ellen Danke (*Tennessean*)
news article and, 126; first formal session
of, 122; Ghost River canoe trip and, 127;
McConnell and author's resolution on,
126; McConnell recommendations and,
127; members of, 122; Stream Obstruction
Removal Guidelines and, 127

Tigrett Technical Committee, skepticism, 110; as
representatives, 110–11; TWRA Nashville
staff skeptics and; 110

Tigrett Wildlife Management Area (WMA),
5; areas south compared to, 58; artificial
swamps at, 59–60; cross levees as highways,
63; channelization contributes to ruin of,
62; declining duck habitat, 58; destruction
caused by, 63; dredging machine at, 62; old
duck club levees and, 60; effects of summer
ponding and, 59–60; floodplains at, 61;
Floyd "Speck" Hurt and author's inspection
of, 59; hunters adapt to, 58; natural
hydrology key to, 61; new management
philosophy tested at, 57; Reelfoot swamps
compared to, 60–61; straight-line channels
and stranded meanders at, **61**; stream
equilibrium and, 62; Yelverton waterfowl
surveys at, 59

Tigrett WMA in 1980: farmer complaints as
catalyst for, 109; flooding complaints at,
109; government help requested at, 109;
hardwood timber at, 109; hunting and
fishing as, 109; Stokes Creek farmers and,
109; 1950s and early 1960s at, 109

thin blue lines, 40, 41, 43, 107

Todd, Robb, 141

traditional (old methods) wildlife management: at
Gooch and Tigrett WMAs, 101;

Twin Rivers Club, 20–22; Unit E, Gooch, 66, 73;
University of Tennessee, 80, 123, 132, 141,
150, 169, 197, 223n13

"unconsolidated muck," 134; university research
and, 132; UTM research and, 132

U.S. Army Corps of Engineers Reelfoot plan:
background for, 159; conflicting role of,
159; county levee boards, role of, 159;
Randy Cook and author objections to,
160; EIS process and, 159; fading Reelfoot
plans and; 158–59; interim water-level
management plan and, 161; Kentucky
farmers opposition and; U.S. Senator Mitch
McConnell and, 161; Myers concessions
and, 160; new spillway held hostage and,
161; reasons to renege and, 161

useless land as best wildlife land: misplaced
tradition, 101; agriculture industry concerns and, 101; government budget
constraints and, 161; wrong policies
and, 161

U.S. Federal Court in Memphis, 150

U.S. Geological Survey, 54, 79, 172, 198

Viar, Doug, 167, 170

Walker, Mary, 151

Walnut Log, **84**, 91, 92, **94**, 155, 158; author's
account of, 91, 92; flooding and, 92; houses
razed at, 158; people at, 91

Ware, Forest, 142

Warr, Ed, 191

Waterfowl hunting, 50, 162, 165; bottomland
hardwoods and, 58; Gooch WMA and,
66, 88; mitigation land and, 97, 98; new
management and, 162; Obion and Forked
Deer rivers and, 19; public land and, 48;
river restoration and, 192

water-level management at Reelfoot Lake, 23,
130, 132, 154, 156–58, 160, 162, 209,
223n14

Water Resources Development Act of 1999, 159,
216

Water Quality Certification Permit (401 permit),
78, 81; Black Swamp and, 193, 194; Tigrett
plan and, 123

Wathen, Greg, 160

Wegner, "Bucky," on Reelfoot drawdown, 142,
143

Wells, Devin, 194

West Tennessee Land Company, 88, 89, 90

report and, 72; Black Swamp issue and, 187, 188; as crippling blow for WTTP, 78; EIS, and, 70; efforts to halt WTTP and, 78; grounds for lawsuit and; NEPA and, 70; injunction and, 70; landowner petitions and, 78; Chester McConnell court appeals and, 74–78; proponents of, 74, 78; Congressman John Tanner and, 78; shutdown of, 78; U.S. Corps of Engineers and Akers lawsuit and, 74–78, 128; U.S. Corps environmental permit withdrawn (May 1989), 78; WTTP environmental impact statement and, 73–74, 159, 160; U.S. Fish and Wildlife Service and, 72
—reformulation of, 216; collapse of, 208, 201; litigants at Black Swamp and, 193; Mission Plan and, 197, 199; Governor Sundquist and, 202; U.S. Congress and, 208; White House proposal and, 202
West Tennessee Tributaries Steering Committee: 215, 216; consensus on Mission Plan and, 198, 199; Chester McConnell and, 126; Gov. Ned Ray McWherter and, 197; Myers and Akers as members of, 205; news media and political support for, 199; Ben Smith as coordinator of, 202; Richard Swaim and, 199; Congressman John Tanner and, 199
West Tennessee Wildlife Resource Association, 169, 221–22n6
Western Energy and Land Use Team (WELUT), 153
wetland ecosystem, 30, 55, 87, 139
wetland management, 50, 103, 174; Mission Plan and, 193; new direction in, 148; new meaning to, 136; northwest territories and, 75; public support of, 210; Reelfoot Lake management and, 144
wetland managers, xix, xxii; challenges to, 47–48; change of philosophy and, 51, 68, 72; continual pressure to appease and, 47–48; domestic crops and, 52; duck clubs and, 51–54; on experience and advice of, 15; history as insight for, 36; "hot crops" versus native crops, 51–52; management duties and, 47–48; public lands and, 47–48; engineers and, 120, 124; experience at Reelfoot and,

162; extreme lake drawdowns and, 139; history as insight for, 36; lack of support for, 50, 183; low-level terraces and, 54; methods and mistakes by, 47; native "moist soils" plants and, 52; new kind of, 48-49; as "duck heads," 48–49; as wetland managers, 51; northwest territories staff members of, 49; new age of management as, 50–51; new waterfowl management methods and, 51–54; nutrient requirements of mallards, 52; scale of one to ten and, 4; short season crops and, 52; "topo" maps as tools for, 41; wetland deficits and, 48
"Wetlands — Common Ground," 172
White Oak Swamp, 121
White Oak WMA, 121, 122
Wildlife Management Institute, 224n15; Chester McConnell and, 72, 75, 76, 77, 124, 193; Gov. Ned Ray McWherter's letter to, 117
Williams, M. V., 116
Williams, Vince, 142
Willoughby, Earl, 11, 33
Wilson, Bobby, 23
Wilson, Steve, 185
Winfred, Doug, 181, 185
Wirwa, Carl, 64, 225n17; Dyersburg public meeting and, 203; Nashville staff meeting and, 191; Tigrett "restoration" team and, 186; as wildlife manager, 49, 174
Wildlife Fund, 173
Wolf River, 3, 127, 170; artificial swamps and, 6; braiding and, 6; cypress swamps in, 5–6; description of, 5; Ghost River and, 5
Wolf River Wildlife Management Area, 6

Yaich, Scott, 48, 219n1
Yelverton, Carl, 59, 65, 999, 214, 220n1

Zerfoss, Frank, xv, 214, 215, 224n15; hope and apathy and, 195; meeting with Myers and Hurst and, 106; mitigation plan and, 103; as northwest territories supervisor and, 97; resignation of, 104; Tigrett duck hunt with, 57, 58; traditional thinking and, 99, 102